Computer Operation for Microscope Photometry

Computer Operation for Microscope Photometry

Howard J. Swatland

Department of Animal and Poultry Science
University of Guelph
Canada

CRC Press
Boca Raton   Boston   London   New York   Washington, D.C.

Acquiring Editor:	Marsha Baker
Project Editor:	Sarah Fortener
Marketing Manager:	Becky McEldowney
Cover Designer:	Dawn Boyd
Manufacturing:	Carol Royal

Library of Congress Cataloging-in-Publication Data

Swatland, H. J., 1944–
    Computer operation for microscope photometry / by Howard J. Swatland
       p.   cm.
    Includes bibliographical references and index.
    ISBN 0-8493-1697-9 (alk. paper)
    1. Microscopy—Data processing.   2. Microphotometry.   I. Title
QH211.S97   1997
502′.8′20285416—dc21   97-36260
                                                                                    CIP

No claim to original U.S. Government works
International Standard Book Number 0-8493-1697-9
Library of Congress Card Number 97-36260
Printed in the United States of America 1 2 3 4 5 6 7 8 9 0
Printed on acid-free paper

Preface

This book describes the general principles of how a personal computer may be used to operate a light microscope for spectrophotometry, fluorometry, polarimetry, and spatial scanning. With the multiplicity of different manufacturers of microscopes and personal computers, it is impossible to give lens-by-lens and wire-by-wire details. This information normally is obtained from manufacturers. But the general principles required for the integration of components, invention of new apparatus, and flexible programming may be obtained from this book.

Being interdisciplinary in scope, the goal is to combine an introduction to light microscopy for personal computer enthusiasts, with an introduction to computer interfacing for microscopists. It is hoped that this will make the subject accessible to readers with little or no experience in either field. Some specific examples of software are provided to illustrate techniques such as armed-card interrupts and stack pointers, but many operations are described as generalized algorithms adaptable to any appropriate high-level language. From a top-down perspective of programming, clusters of similar operations are identified and described, particularly the control of scanning devices and photometers for step-and-measure operations. This allows a generalized control program to cope with almost any configuration of the microscope. Simple, inexpensive methods for computer control of the microscope are considered, together with the basic principles of relatively expensive commercial systems. Suggestions are made for liberating commercial equipment from its often restrictive software.

Chapter 1 introduces some of the basic features of light microscopy and personal computer interfacing, such as Köhler illumination and the general purpose interface bus. Chapter 2 examines illuminators: how they may be programmed and how they affect system performance and software. Microscope spectrophotometry, for example, requires particular attention to the emission spectrum of the source. Chapter 3 completes the topic of light intensity control by considering how light may be turned on and off using shutters and apertures. Chapter 4 considers how to make automated measurements, using a variety of different photometers ranging from inexpensive photodetectors to high-voltage photomultipliers.

Having reached the point of being able to make photometric measurements, Chapter 5 deals with spectrophotometry, using different types of monochromators to measure spectra. From the perspective of writing universal software, however, there is not much difference between incrementing a monochromator and making a measurement, and incrementing a scanning stage and making a measurement. Chapter 6, on stepper motors and servomotors, confirms the universality of step-and-measure programming and shows how the same software used for spectral scanning in spectrophotometry also may be used for spatial scanning.

Chapter 7 considers polarized light, a routine tool in optical mineralogy but very neglected in biological light microscopy. Step-and-measure software now is used to automate analyzer measurements in a polarized light microscope. Chapter 8 augments the microscope hardware by adding epi-condenser barrier filters and dichroics and deals with special problems occurring as we move from microscope spectrophotometry to microscope spectrofluorometry.

Chapter 9 introduces some of the basic principles of video image analysis, using a frame grabber to extract small data sets to manipulate for special applications. Chapters 10 and 11 relate to photodiode-array spectrographs and fiberoptics and to techniques that allow us to liberate the light microscope from the laboratory bench and move it out into the real world. Many of the operations undertaken on parts of an image also may be undertaken on parts of a remote specimen. Thus, if a sample is too big to put on the microscope stage, an optical fiber may be used to bring the remote specimen to the optical axis of the microscope. Finally, Chapter 12 considers how the personal computer may be used to put a microscope sample through a series of programmed changes in pH, temperature, or fluid environment, so that the whole assembly of computer, microscope, and support system becomes a powerful tool for automated experimentation.

There are many excellent textbooks covering the theory of light microscopy and personal computers. Building on these foundations, the objective here is to describe their powerful hybrid, the computer-assisted light microscope.

The support and encouragement of editor Marsha Baker are warmly acknowledged, as is the friendly efficiency of Sarah Nicely Fortener as project manager. I much appreciate Bridget O'Brien gleefully ferreting out many of the numerous errors in my manuscript. And, to the wonderful people at Hewlett-Packard and Carl Zeiss, thanks for answering all my technical questions so politely over the years. I doubt whether the material I am presenting is perfectly free from errors, which are all my fault and for which I apologize in advance. When they are detected, the errors will be listed with all my other embarrassments on my home page, accessible via **http://www.aps.uoguelph.ca/swatland/html/gasman.html**.

The Author

Dr. Howard J. Swatland obtained his B.Sc. from the University of London (1967), and M.S. (1970) and Ph.D. (1971) from the University of Wisconsin, Madison, working on the prenatal development and innervation of skeletal muscle. After a postdoctoral fellowship with the Muscular Dystrophy Association of America and a senior fellowship in pathology at the Central Veterinary Laboratory, Weybridge, England, he joined the faculty at the University of Guelph in 1974. He assembled his first microscope from surplus parts in 1960 and his first analog photometer in 1970, and started programming and interfacing with a Motorola 6808 in 1981. His initial applications of microphotometry were for quantitative histochemistry of muscle growth and metabolism, which later evolved into the direct investigation of tissue structure and composition using fiberoptics. He has authored 184 refereed scientific publications, three textbooks on animal growth and composition, and seven chapters for other books. He has received four awards for teaching and a research award from the American Society of Animal Science. However, as a single parent raising two children, his proudest achievements are in the domestic sciences, with honors in French braiding for his daughter, winter camping with his son, and starched linen at Sunday lunch.

Contents

Chapter 1

Introduction

1.1   Purpose of this Book

Very few of the millions of ordinary light microscopes (LMs) in everyday use are connected to any of the millions of ordinary personal computers (PCs) in everyday use, even though the distance between most LMs and their nearest PC may be less than the length of a long printer-cable. This is curious, because the possibilities of combining the LM and PC have been evident since the late 1970s (Rasch and Rasch, 1979), particularly for microscope spectrophotometry, fluorometry, or polarimetry or for operations involving scanning or tilting a sample. Thus, instead of gleaning thousands of laborious measurements from the LM, transferring them to laboratory notebooks, and then entering them on a computer for analysis, data may be collected and analyzed automatically.

Computers are essential for many newer types of light microscopy, such as video and laser confocal microscopy, but these are well documented in our libraries (Inoué, 1986; Chen et al., 1994; Pawley, 1990; Lewis, 1991; Shotton, 1991). As may be seen from the current literature on microscopy (Mannheimer, 1996), however, much less information is available on computer operation of the ordinary LM. Hence, the purpose of this book is to describe how an ordinary LM, one not involving a flying spot or heterodyning (Samekh, 1990), can be coupled to a PC to produce what we will call, for the sake of simplicity, a computer-assisted light microscope (CAM).

This book cannot substitute for a manufacturer's specific instructions for CAM components but rather aims to provide background information and general ideas. Much of the software for an off-the-shelf commercial CAM is limited in scope and cannot cope with any change in the optical layout of the system. It may force the CAM to become a machine, always doing the same thing. On the other hand, home-grown custom software may liberate the CAM, allowing flexibility and adaptability and encouraging expansion into new fields. Thus, users must program their own systems to gain full advantage of the CAM as a programmable optical robot.

Fortunately, the programming skill required is nothing special, as may be seen from the subprograms included in this book as examples.

When presenting examples, the following format is used for parentheses: no parentheses for a line number from a program — e.g., 1000; curly brackets for a variable from an array — e.g., {9}; square brackets for a step from a generalized algorithm — e.g., [10]; and curved brackets for an item listed in the text — e.g., (5).

There is a great tradition of inventiveness among light microscopists, who often assemble their own CAMs from surplus components. Thus, recycling and adapting components are laudable. The performance of a CAM built from odd and recycled LM parts can, with a modern PC as controller, rival that of a very expensive CAM of the 1970s (David and Galbraith, 1974; Galbraith et al., 1975). Potentially useful components on an older LM may be identified from Hartley (1993), using Pluta (1988) for their theory of operation. It would, however, be criminal to alter or adapt anything from the brass microscope era, which should be treasured for its historical value. It is much easier to set up a CAM with modern, infinity-corrected optics than it is with older, 160-mm tube-length components requiring Telan lenses to correct for lengthening of the tube. It is hoped that the purpose of this book, which is to show how easy it is to build and operate a CAM, will be achieved if readers new to the field are encouraged to get started.

1.2   Power Supply

1.2.1   Stabilization

A 12-V halogen lamp operated at 100 W is adequate for making measurements of transmitted and reflected visible (VIS) and near-infrared (NIR) light, but may be problematically weak around 400 nm and almost certainly inadequate farther into the ultraviolet (UV), where the lower limit available for the LM is about 220 nm (Wood and Goring, 1974). A stabilized DC power supply is highly desirable; otherwise, much time will be spent in integrating photometric measurements to compensate for power-supply ripples. Often it is instructive to examine the DC power source on an oscilloscope: the DC output may be good, but the connections may be bad, or vice versa. In older CAMs, sometimes it was necessary to solder the power cables directly to the lamp, rather than relying on spring-loaded contacts in the lamp housing. Two-pin halogen lamps with a tight fit in a ceramic base may be more reliable in this regard. But, even here, depending on the way in which the halogen lamp is mounted, it may be possible for a broken filament to dangle, complete a circuit, and dance a small minuet to create random fluctuations in light intensity. Lamps should be replaced regularly, and contacts should be burnished with a brass brush. An off-the-shelf, general purpose, 0- to 20-V, 0- to 10-A power supply that is both stabilized and programmable is definitely worth investigating and may be less expensive and far more versatile than a commercial CAM power supply.

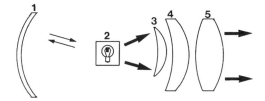

FIGURE 1.1

Illuminator components: reflector, 1; lamp, 2; and collector lenses, 3, 4, and 5. Narrow arrows show the action of the reflector, and broad arrows show the collimation of the output beam.

1.2.2   Arc Lamps

Arc lamps are essential for UV as well as providing useful peaks for VIS and NIR, but building a power supply for the Osram short-arc mercury (HBO) and xenon (XBO) lamps widely used for UV and fluorescence microscopy is best left to the professional electrical engineer. Start-up voltages for ignition are dangerously high, meaning that careful attention should be given to the state of cable insulation and grounding for any parts recycled from elsewhere. Surge protection for a PC plus a serious level of cable shielding may be required if an arc lamp is to be used anywhere near a PC. If you hear a PC printer reset itself when an arc lamp is ignited, you have a problem that should be fixed before it is too late.

1.3   Illuminator

Most research LMs have a detachable lamp housing for the illuminator and may even have two or more illuminators switched in or out of the light path by mirrors. Since a solenoid shutter in front of the illuminator also is required for a CAM, there are many reasons why a CAM may remain dark when turned on. Some solenoid shutters automatically swing to the open position when turned off, while others do not. Thus, a power-up routine should be included for such items so that, when a CAM is first turned on, the user can at least start with a working microscope. Someone else using the CAM in manual mode or with different software still may be left in the dark, however, so an equivalent power-down routine is needed when quitting the program, leaving the microscope in a workable state for other users.

1.3.1   Adjustment

Figure 1.1 shows a typical layout for the parts of a halogen illuminator. The reflector adjustment should be checked, especially after replacing the lamp. There are usually three reflector adjustment screws in the lamp housing: up and down, left and right, and for the lamp-reflector distance. The centering and focus of the reflector may be examined by removing the illuminator and aiming it at a wall. After the reflector orientation and focus have been adjusted to superimpose both the primary bright image of the lamp and its slightly less bright reflection, the adjustment is altered by moving the reflected image slightly to the right or left of the primary image. This produces a larger, more uniform field of illumination and may help to avoid overheating the lamp by reflecting heat back to its source. When using a lamp off its microscope, however, it is essential to treat it gently, keeping it level. A powerful lamp can be blinding and can set fire to curtains (as discovered by the decorators who burned down Windsor Castle). A critical point about some types of arc lamp is that they should only be operated in a vertical position. Tilting of the arc can lead to disaster or to a shortened life expectancy for the burner. Adjustment of the lamp when installed on the CAM is done by removing the substage condenser, racking the condenser carrier to its highest position, and using it to support (temporarily!) a sheet of tracing paper on which the direct and reflected images of the lamp can be adjusted. Some epi-fluorescence microscopes may have a built-in screen for this purpose.

1.3.2   Assembly

The shape and layout of the collector lenses in Figure 1.1 are only diagrammatic but provide a guide to the reassembly of a three-lens system if it has been dismantled. One, two, or three lenses may be present, depending on the type of illuminator. Make careful notes when dismantling any optical system for cleaning, because the components often can be assembled in a myriad of different ways, only one of which is correct. The distance between the lamp and the collector lenses may be adjusted to collimate the output beam. Should alterations or substitutions be made to a lamp housing, it is essential to ensure that the first collector cannot get too close to the lamp, especially in the case of an expensive aspheric quartz collector lens relative to an arc lamp. Even when used carefully, such items are easily cracked by heat.

If a lamp housing is dismantled, take a cautious look at the reflector, which may be silvered on its front surface (not like an ordinary mirror where the silvering is behind the glass). A front-silvered mirror will be scratched by even the most gentle contact with a lens paper, and horribly polluted by a finger print. Ultrasonic cleaning is required. The tightness of fit of collector lenses in their supporting rings also should be noted, because the lenses may get cracked by heat if they are reassembled too loosely or too tightly. The glass surface of a new lamp should never be contaminated by finger contact; otherwise, the carbonized fingerprint will absorb heat and may lead to early failure of the lamp. Ethanol may be used to clean a dirty lamp. Above all else, remove and replace arc lamps as you would the detonator of a miniature hand grenade, wearing leather gloves and protection for face and arms. A high-pressure lamp with brittle glass from over-use is dangerous, not to mention the potential for mercury contamination. Finally, when a lamp housing is reassembled, check to ensure that the swing-in diffusion disk is where you want it. It should be in the optical pathway for general microscopy, but out of the pathway for alignments or when low light levels pose a measuring problem. For built-in illuminators, access to the swing-in diffusion disk may be cunningly concealed somewhere along the light tube under the base of the LM.

1.4   Köhler Illumination

1.4.1   Lamp Size

Finding that the LM has a swing-in diffusion disk begs an obvious question: why is it there? In the early days of microscopy, obtaining sufficient light for sample illumination sometimes was a problem. If sunlight was available, it was directed into the optical axis of the LM with a flat mirror. When relatively large electric lamps became available for domestic use, light was directed into the optical axis of the LM using a concave mirror, ideally with the frosted inner surface of the lamp bulb at the focal point of the mirror. This is why LMs without built-in illuminators have a two-sided (flat and concave) mirror at the bottom of the optical axis. Using light from a frosted domestic electric light bulb (or a ground glass plate illuminated from behind by a clear lamp), the substage condenser was focused so that the frosting or ground glass was seen in the same plane as the specimen. This method of illumination using a large lamp is called critical illumination and is quite adequate for many aspects of ordinary microscopy, but it may lead to chromatic errors in spectrophotometry (Florijn et al., 1996). Once automobile head-lamps became available, Köhler illumination became easy to use. Köhler illumination is the best method for microscope spectrophotometry, as explained by Piller (1977).

1.4.2   Adjustments

Figure 1.2 introduces the practically important features of Köhler illumination. With the specimen mounted beneath a cover slip on a microscope slide (Figure 1.2, 1) and supported on the stage of the LM (Figure 1.2, 2), we normally have control of four features of the illumination pathway: the height of the substage condenser (Figure 1.2, 3), the diameter of the condenser aperture (Figure 1.2, 4), the diameter of the field stop (Figure 1.2, 5), and the brightness of the light source (Figure 1.2, 6). Figure 1.2 is a gross simplification of both image-forming and illuminating ray paths and completely ignores wave optics. It does not show that the illuminating rays are essentially parallel as they pass through the specimen, thereby minimizing any effects of inhomogeneity of the source.

For a vintage LM with a mirror and no built-in illuminator, step 1 of establishing Köhler illumination consists of getting the LM illuminator directly in line with the LM (not off to one side), getting the bright spot from the illuminator in the exact

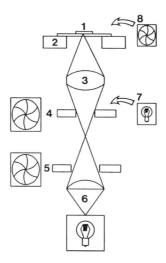

FIGURE 1.2

Köhler illumination: specimen beneath a cover slip on a microscope slide, 1; microscope stage, 2; condenser, 3; condenser iris aperture, 4; field stop iris, 5; illuminator, 6; image of the illuminator filament at the level of the condenser aperture, 7; and image of the field stop at the level of the specimen, 8.

center of the mirror, then rotating the mirror so the bright spot centers on the substage condenser. With a properly adjusted built-in illuminator, this step already has been made, and light from the illuminator should pass into the optical axis of the condenser. This may be confirmed by following the bright spot of the illumination beam on a sheet of tracing paper. If the bright spot does not disappear into the optical axis of the condenser, then check the centering of the illuminator and the lenses and mirrors beneath the LM. An old LM unlovingly cast onto a concrete floor in a basement storage area may well have sustained damage to this vital part of its optical system, which often is unprotected beneath the stand of the LM. Even for a new LM, another indirect source of error is dust, which sooner or later prompts the least talented person of a research team to attempt to clean the well in the LM stand whence issues the light. Reassembling an optical system is never as easy as taking it apart.

Further centering of the illuminator is checked as follows. With the swing-in diffusion disk out of the light path, with a mounted specimen on a microscope slide already in focus with a lower power (×10) objective, and with the condenser aperture wide open, the condenser is racked up carefully to its highest point. In most cases, there will be a mechanical stop to prevent the top lens of the condenser from crashing into the underneath of the microscope slide. The stop is adjustable, though, and may have been removed altogether, particularly if an earlier user changed the condenser for some reason, so care is needed the first time. The best condenser for microphotometry is, in fact, another microscope objective with a magnification and numerical aperture one step lower than that of the actual objective in use. Blindly racking up an objective used as a condenser may lead to disaster long before the mechanical stop is engaged. Thus, for becoming familiar with Köhler illumination, it might be wise to substitute an ordinary condenser. Interchanging and centering condensers is a repetitive operation for many types of microphotometry and is another skill to be learned until it becomes automatic. Finally, with a condenser racked to its upper limit, remove one ocular (eyepiece) from the LM to observe the back lens of the objective, which should be evenly illuminated. If it is not, re-adjust the illumination pathway again. Replace the ocular and focus the collector lens of the illuminator so that the filament or arc can be seen clearly. If an arc lamp is used, this is a good point at which to confirm that the primary image and its reflection are side by side to form a neat rectangle of brightness (provided a UV barrier filter is in place).

Step 2 for Köhler illumination is to close the field stop down to a small aperture, then to lower the condenser (already at its highest position) until the field stop aperture is sharply in focus (the specimen is already in focus). The field stop aperture then is centered using the centering screws of the condenser or condenser carrier. If there is more than one set of centering screws, swing out or remove the unit with the secondary (high magnification) set, center the coarse or low magnification components, then replace and re-center the secondary set. With the field stop centered and in focus, open the field stop until it is just out of sight. If the complete field shows any sign of uneven illumination, now is the time to make fine adjustments to the illuminator.

The final step is to reduce the condenser aperture to a reasonable diameter — which of course means nothing to the first-time user. So, for a rule of thumb that is good enough for most normal situations, remove the microscope ocular again. Close the condenser aperture until it has reached the outer third of a radius of the visible field, then replace the ocular. Closing the condenser aperture increases the contrast but decreases the resolution. Thus, the ideal position depends on the natural contrast of the specimen and on how much detail is needed. Beginners tend to err on the side of too much contrast, setting the condenser aperture too small, while other users may make the opposite error, sacrificing contrast and glare reduction by opening the condenser aperture too wide. This may occur when, prompted by a poor electronic signal-to-noise ratio, the user attempts to strengthen the electronic signal by opening the condenser aperture. This introduces a steady background of glare — light that does not contribute to the image. The condenser aperture also controls the depth of focus in the specimen.

1.4.3   Testing

With Köhler illumination properly established, a sheet of tracing paper should reveal that there is an image of the lamp filament at the level of the condenser aperture (Figure 1.2, 7). Any time the field stop is reduced, it should appear in focus around the specimen (Figure 1.2, 8). Thus, the field stop corresponds to the fixed aperture of the objective, so that sensitive specimens are not subjected to unnecessary heat and light. Filters to remove heat by reflection or absorption are generally located somewhere along the light path before the specimen, but they may not be very effective or they may have been removed. By some unfathomable law of microscopy, only good specimens are damaged by heat; poor specimens seem to have a natural immunity.

1.4.4   Problems

It is important to become familiar with Köhler illumination so that it becomes instinctive when using the LM; however, having stressed the virtues of optimum alignment and illumination, it must be admitted that sometimes it is difficult to establish Köhler illumination properly. In many interesting experiments, such as following real-time changes in a specimen during computer-controlled alterations of its temperature, pH, orientation, or surrounding environment, we may be forced to abandon standard thickness microscope slides (1.1 ± 0.1 mm thick with a refractive index of 1.52 ± 0.01 at 546 nm) and cover slips (0.17 mm thick) for which ordinary LM optics are designed. Thus, as slides, cover slips, and sample media get thicker, the image of the field stop becomes progressively less distinct. For ordinary viewing this may not be a problem, and the field stop is out of view anyway, but it can become a problem when swing-in fixed field stops are used for microphotometry. The problem is imperceptible at low magnifications, using objectives with apertures < 0.3, but it soon appears at higher numerical apertures where cover glass thickness deviations of even 0.01 mm become detectable at the highest apertures. This effect sometimes limits the resolution possible with an experimental chamber. Perfectly prepared conventional microscope slides are necessary for maximum resolution with high magnification. Ingenious sample chambers for observing specimens during computer-controlled experiments may be limited to intermediate ($\times40$) or low magnification, unless the sample can be brought into contact with the lower surface of a thin cover slip.

1.5   Resolution

There is no upper limit to the magnification attainable with the LM, and it is quite acceptable to project an over-magnified image so that its contours or some other morphometric feature can be identified with a computer stylus or mouse. Thus, hand-held errors are minimized relative to the overall image, but as soon as any highly magnified image is examined critically, it becomes obvious that there is a limit to the useful magnification attainable. Beyond the useful limit the image may be larger, but it is more blurred and reveals nothing that cannot be seen more clearly at a lower magnification. Sometimes it is possible to see things in an over-magnified image which have no existence in the actual specimen — such as granular effects caused by diffraction. Thus, resolution is the ability to form separate images of two points close together. Empty magnification should be used only for convenience in measuring, not for directly investigating the structure of the specimen.

The design of the LM objective has a dramatic effect on resolution. When an image is formed it is surrounded by a series of interference fringes. An objective accepting a large angular cone of light from a specimen has a higher resolving power than an objective only accepting a small cone. For example, in the case of a sheet of aluminum foil with a very small pin hole, the image is a bright spot encircled by a halo of reduced brightness, the Airy disk. If there are two such pin holes close together, acceptance of a large cone of light may enable the central bright spot of the second pin hole to be seen separately just inside the Airy disk of the first hole. It may then be possible to observe that there are two separate round pinholes, both encircled with an Airy disk, not one elongated oval hole in the specimen, as seen with a low aperture objective.

1.5.1   Numerical Aperture

The angular size of the cone of light that can be accepted is given by the numerical aperture of the objective:

$$\text{numerical aperture} = N \sin U$$

when N is the refractive index in the space around the specimen (for practical purposes, below the slide and above the cover slip), and U is the angle in the axis of the cone of light. According to Piller (1977), the smallest resolvable distance (d) in μm is given by Abbe's theory of image formation as:

$$d = \lambda/(NA_{Obj} + NA_{Con})$$

where wavelength, λ, is expressed in μm, and values of the numerical aperture (NA) are for the objective (Obj) and condenser (Con). Thus, with the LM, there are three ways to increase the resolving power: (1) decreasing the wavelength, (2) increasing the size of the angular cone of light towards its theoretical maximum of 90°, and (3) increasing the refractive index in the object space. In Figure 1.3, for example, without immersion oil to link the top of the cover slip to the front lens of the objective, only a small aperture angle is possible (Figure 1.3, 1) relative to that possible with oil (Figure 1.3, 2). With the dry objective, the limitation to the aperture is caused by internal reflection at the top of the cover slip (Figure 1.3, 3), which is governed by the refractive index of the medium in contact with the cover slip. This principle of optical linking is exploited elsewhere, as in linking the top of the condenser to the bottom of the microscope slide and in working with optical fibers.

1.5.2   Comparison with Electron Microscope

In an electron microscope (EM), a stream of electrons from a tungsten filament is manipulated with magnetic lenses formed by a flat electrical coil in a hollow iron ring. The electrons are deflected through a spiral pathway so that, in its overall configuration, the system functions as a compound microscope. The effective wavelength of the electron beam is dependent on the voltage at which the electrons are

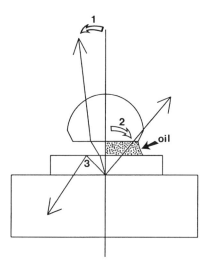

FIGURE 1.3

Immersion oil increasing the aperture of an objective: low aperture with a dry lens, 1; high aperture with oil linking the cover slip and objective lens, 2; and rays lost by internal reflection with a dry lens, 3.

accelerated but is always far smaller than in the LM. Values around 5 pm are typical; however, this tremendous gain is offset by ordinary magnetic lenses in an EM having a very low numerical aperture (<0.01). The resolving power that results from this combination is typically about 1.5 nm for a biological specimen. However, although resolution is extremely important in both light and electron microscopy, it is often the method of preparing the specimen for examination that poses the actual limitation of resolution, rather than the microscope itself. Thus, high-quality modern specimens examined in an antique LM generally appear far better than specimens prepared by ancient methods examined in a high-quality modern LM (Bracegirdle, 1978).

1.6   Microscope Objectives

Harnessing the full potential of the CAM encourages us to innovate and experiment with different ways of interfacing the LM with the sample, but there is not much that the average user can do with the microscope objective, except to appreciate its properties and use it properly. High achievers who wish to make, adapt, or test their own objectives should consult Orford and Lockett (1931). Most of the objectives that can be liberated from an older research LM are based on a tube length of 160 mm. Thus, if the tube length is increased by inserting a computer-operated device of some sort into the light path, Telan correction lenses are needed before and after the insertion. These may decrease overall transmittance and reduce image quality, although the effect may be difficult to detect in normal situations. Infinity corrected optics, long available from American Optical but now available from all research LM manufacturers, avoid this problem by using parallel light rays through the barrel of the microscope which, thus, no longer needs to be fixed in length at 160 mm.

1.6.1   Achromats

The chromatic aberration of a simple lens (blue light coming into focus closer to the lens than red light) is canceled by combining a strong positive lens of crown glass with a weak negative lens of flint glass. The latter has a higher refractive index than the former, bringing red and blue into focus at the same point but leaving the lens with an overall positive effect. This correction, the classical achromatic objective, still leaves blue plus red out of focus with green, but the worst of the problem has disappeared. However, we now have a compound lens composed of different types of glass glued together, which raises an important practical point.

1.6.2   Cleaning

In everyday use, it is not long before the front lens of the objective needs cleaning, either to remove traces of old immersion oil or the odd fingerprint. In most laboratories, LMs are surrounded by a tempting variety of strong solvents such as xylene and ethanol. These are excellent for cleaning dirty glass, but they are also excellent at dissolving the cement between lenses. Thus, strong solvents carried into the objective by capillary action soon erode both the cement and the quality of the lens. A safe cleaning solution is something like 65% methyl acetate, 30% ethanol, and 5% diethyl ether, but even this is only used after first attempting to clean the objective in other ways. Fingerprints and old immersion oil are easily removed with new immersion oil on lens tissue. Finish up with the time-honored method of "Haa!". Close the back of the objective with your thumb, put the front lens up to your open mouth, then go "Haa!". Clean off the condensation and polish the lens carefully with soft lens tissue.

1.6.3   Enhanced Achromats

Most achromatic lenses are also aplanatic, corrected for spherical aberration; otherwise, paraxial rays coming through the outer circumference of the lens aperture would tend to come into focus closer to the lens than those coming through the central axis of the lens. The human eye is very sensitive to green light, so aberration for green light is canceled by balancing the relative curvatures of successive lens in the objective. The result is a widely available, relatively inexpensive lens of considerable interest to the experimenter because it is robust and has a large working distance (the gap between the front lens and the cover slip). Some experiments involve a certain risk to the objective lens if the sample is heated or moved vertically, in which case, the first choice is an inexpensive achromat. Thus, it is useful to know that ordinary achromatics can give perfect results with monochromatic light (focusing at one plane) and a small measuring aperture (focusing only at the center of the field). Achromats at higher cost and complexity can be obtained with increasing degrees of flat-field correction so that the whole field of view is in focus at once, not just the central region. With full flat-field correction the objective is called a planachromat. Achromats also can be obtained with a long or very long working distance (such as the Zeiss LD planachromat and UD achromat series). There is a loss in numerical aperture but they can focus on a distant sample maintained in a specially built chamber, which opens up a whole realm of dynamic CAM experiments.

1.6.4   Apochromats

Research LMs often are equipped with apochromatic objectives and compensating oculars. To eliminate the difference in depth of focus of green vs. red plus blue, all three colors are brought into focus in the same plane, with the inevitable consequence that the images are slightly different in size. This is corrected by the compensating ocular, which equalizes the image sizes. This also makes it possible to correct for spherical aberration at all three wavelengths, not just for the dominant green light, as in an achromat. The semi-apochromat is a variant of the apochromat which gives close to the same result without including a fluorite lens, as in the Zeiss originals. In Nikon CF achromats, the computer-calculated corrections are so good that the CF achromats rival older apochromats in performance but do not require matching compensating oculars.

1.6.5   Fluorite Objectives

The performance of the apochromat also is rivaled by other objectives such as neofluars which use fluorite as the main lens material, not just for correction. Neofluars have a high numerical aperture and a high transmittance of UV light, produce very vivid images, and can be corrected to give a flat field (plan-neofluars). Neofluars or plan-neofluars are essential for UV microscopy and are the first choice for fluorescence microscopy. Although they fall slightly short of the maximum resolution available with apochromats, the difference is seldom important except for those in pursuit of the finest details resolvable by light microscopy. Plan-neofluars also are available as immersion lenses at low magnifications (×16 to ×40) with a most useful feature: they can be immersed directly into water or glycerin for the direct observation of living samples. Ultrafluars are special purpose objectives corrected for the thick (0.35 mm) quartz glass cover slips which are necessary for UV absorbance measurements and are very useful in the analysis of pulp and paper fibers (Morrison et al., 1996). Although they are achromatic over a wide range of UV as well as visible light and, hence, are ideal for LM spectrophotometry, they tend to be very sensitive to changes in temperature.

FIGURE 1.4

An epiplan POL objective mounted in a centering ring for direct attachment to an epi-condenser (1) and a POL objective with built-in centering rings (2).

1.6.6   Polarized Light Objectives

At some point in the manufacture of a glass lens, the molten glass must be cooled and solidified. This can introduce stress lines that interact with polarized light. Thus, for polarized light microscopy, stress-free objectives are required, usually POL achromats, planachromats, or neofluars. Sometimes, though, ordinary objectives are stress free and can be used for polarized light microscopy after extensive checking, which usually involves rotating the objective and measuring its transmittance through parallel polarizers before and after the objective. However, if centering the objective in the LM axis is critical, the only option may be to use a POL objective with centering rings (rotating rings on the barrel of the POL objective; Figure 1.4, 2). The rotating rings on a POL objective should not be confused with the rotating correction collars found on top-quality objectives which are for adjusting the objective to different cover-glass thicknesses. Less often, one may encounter a high-aperture objective with its own iris diaphragm operated from a rotating ring on the barrel. This is for matching the objective to the condenser for dark-field illumination.

1.6.7   Epi-Objectives

In epi-condensers for reflected light microscopy of solid samples, a central axial beam of illumination is directed downwards through the objective by a beam splitter in the illuminator, which is now located above, rather than below, the objective

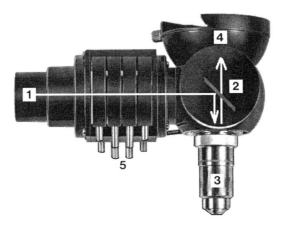

FIGURE 1.5
The principle of the epi-condenser. Light from the illuminator (1) is reflected downwards by a beam splitter (2) into the objective (3). Light from the sample passes through the beam splitter into the optical axis of the LM (4). The beam splitter may be a half-silvered or dichroic mirror, a prism, or a split mirror. This condenser is from an old epi-fluorescence microscope and contains heat and exciter filters (5); however, a field aperture for photometry also could be located near position 5.

(Figure 1.5). Thus, there may not be much space available above the specimen (because of the presence of the epi-illuminator), and epi-objectives may be short and mounted individually in centering rings that connect directly to the epi-illuminator (Figure 1.4, 1), thus saving space by not having a revolving nosepiece to hold all the objectives parfocally. If a revolving nosepiece is fitted, the epi-objective may have a cylindrical collar to increase its length to that of a typical objective. Another configuration is to use a hollow cone of illumination outside the objective, as in the Leitz Ultropak, but the objective then may have a very large diameter. Long-working-distance epiplan objectives (LD-epiplan) are very useful for a CAM, and they make ideal substage condensers for transmitted light photometry. Epi-objectives are routinely used in materials science and have many novel applications, such as microcolorimetry (Piller, 1973), measuring microscopic curvatures (Stavenga and Leertouwer, 1990), in combination with micellar electrokinetic capillary chromatography for porphyrins (Yao and Li, 1996), and in combination with pulsed field gel electrophoresis to detect individual molecules of DNA (Gurrieri et al., 1996).

An epi-objective does not usually require a cover slip, because it is intended for use with reflected light on a solid sample. Although one might initially think this to be a good way of examining biological tissues, the drawback is that most biological tissues are approximately 80% water, and the front surface reflection from wet tissues produces a very bright glare. There are several solutions to this problem, such as using a quarter-wave plate on the front lens of the objective (to produce elliptically polarized light unable to re-enter the quarter-wave plate in the reverse direction), a polarization block consisting of two crossed polarizers and a plane glass reflector, or a central stop in front the illuminator collector lens (Cornelese-ten Velde et al., 1990).

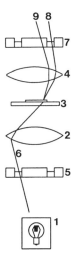

FIGURE 1.6

Principle of phase contrast: illuminator, 1; condenser, 2; specimen, 3; objective, 4; phase annulus, 5; illuminating ray, 6; phase plate, 7; direct ray, 8; and diffracted ray, 9.

The extra internal reflections produced by some of these components may be minimized by finding an appropriate thickness of electrical insulating tape to include on one side of the mounting, thus diverting unwanted reflections off-axis.

1.6.8 Phase Contrast

The image in an ordinary LM is created by differences in the amplitude of light corresponding to those of the specimen, whereas the image produced by Zernike's phase contrast method originates from differences in the phase of the light waves. Thus, natural interference from differences in refractive index or thickness of parts of the specimen is converted to visible patterns of destructive interference (Ross, 1967).

Figure 1.6 is a greatly simplified diagram to show the principle of phase contrast. In Figure 1.6, the illuminator, condenser, specimen, and objective are indicated by parts 1 to 4, respectively. For transmitted light, a phase annulus or ring is located below the condenser (Figure 1.6, 5). It blocks the light around the edge and in the center, only letting through a ring of light (Figure 1.6, 6). Near to the back-focal plane of the objective is located a phase plate on which has been deposited a ring of metallic material which partially absorbs transmitted light (Figure 1.6, 7). Direct rays, which have passed through the phase annulus, through the specimen without diffraction, and through the ring of the phase plate (Figure 1.6, 8), interfere with rays that have passed through the phase annulus, were diffracted by the specimen, and just missed the ring of the phase plate (Figure 1.6, 9).

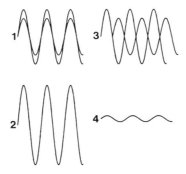

FIGURE 1.7
Principle of constructive and destructive interference. If two waves are in phase (1), constructive interference increases amplitude to appear bright (2). If two waves are out of phase (3), destructive interferences reduces amplitude to appear dark (4).

Whether the direct and diffracted rays interfere constructively or destructively (Figure 1.7) depends on whether the ring of the phase plate advances or retards the phase of the rays passing through it. To advance the phase, the phase plate ring is a groove (to give a short light path through the phase plate, as in Figure 1.6, 8). To retard the phase, the ring is elevated above the plate (to give a long light path through the phase plate). Most phase contrast objectives are positive, so that thick or highly refractive regions of the specimen appear dark on a light background.

A typical positive phase contrast system generally produces haloes around parts of the specimen that are thick or have a high refractive index, with a bright halo around the outside and a dark halo on the inside of the structure in question. Haloes are caused by some of the diffracted light reaching the ring of the phase plate, depending on the size of structure and its refractive index contrast with its surroundings. A first approach to reducing haloes if they become a problem is to change the magnification (and hence the size of the structure relative to the phase annulus). A second approach is to alter the refractive index of the mounting medium around the structure, if this can be done without causing the specimen to wilt or burst by osmosis.

Although the same phase annulus can be used for positive and negative phase contrast objectives, the annulus and phase plate must be matched in size for different magnification objectives. Thus, the condenser of a phase contrast LM has a series of different sized phase annuli which are rotated into the light path to match the objective in use, although several different low magnification objectives may share the same annulus (Figure 1.8). The centering of the annulus relative to the phase plate must be checked regularly by examining the back of the objective, either with a telescope that temporarily replaces the ocular or by making the ocular function as a telescope by swinging in a built-in Amici-Bertrand lens below the ocular (Figure 1.9). Somewhere on the condenser are adjustments to center the annulus, although they are sometimes hard to find and may require a special tool. Focusing telescopes and Amici-Bertrand lenses are particularly useful in LM spectrophotometry because they are ideal for many other types of centering adjustments.

FIGURE 1.8

A substage condenser for phase contrast, showing the optical axis (1) and the casing around a revolving disc of different phase contrast annuli (2).

1.6.9   Differential Interference-Contrast

Nomarski's method of differential interference-contrast (DIC) has proved the most successful method of interference microscopy in general use, although it is does not lend itself readily to quantification. This requires dual beam interferometry, such as in Horn and Jamin-Lebedeff microscopes (Dunn, 1991). In essence, DIC functions like a split-beam interferometer, in which the beam passing through the sample interferes with a reference beam bypassing the sample, except that, to avoid the complex engineering necessary to allow the reference beam to bypass the sample, the reference beam also passes through the sample but with a slight offset and in a different plane of polarization. This is achieved by using a Wollaston prism as beam splitter (Figure 1.10).

A Wollaston prism consists of two birefringent quartz prisms cemented together with their optical axes perpendicular. Plane polarized light produced by a polarizer at 45° (Figure 1.10, 1) splits into two polarized rays in the lower part of the Wollaston prism (not seen in Figure 1.10, 2, because one is behind the other). The separation is enhanced in the upper part of the Wollaston prism, with ordinary and extraordinary rays now diverging at a greater angle and polarized at 90° to each other. After passing through the condenser, specimen, and objective (Figure 1.10; 3, 4, and 5), the beams

FIGURE 1.9

Properly (1) and improperly (2) centered phase annuli.

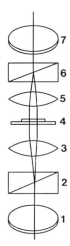

FIGURE 1.10

Principle of Nomarski differential interference-contrast: polarizer at 45°, 1; Wollaston prism, 2; condenser, 3; specimen, 4; objective, 5; adjustable Wollaston prism, 6; and polarizer at 135°, 7.

are recombined by another Wollaston prism (Figure 1.10, 6) and passed through a polarizer at 135° (Figure 1.10, 7). Interference across the field of view, corresponding to the separation of the two polarized beams passing through the specimen, produces a shadowing effect. But, of course, if the specimen is rotated, the shadow-like regions of destructive interference will change in position and often in appearance. Thus, care is necessary in standardizing the orientation of specimens such as fibers, crystals, and linear striations which may be optically active themselves.

In most DIC systems, the magnitude of interference is adjusted by sliding the upper Wollaston prism across the optical axis to change the relative depths of its two component wedges. A DIC system uses a normal condenser iris aperture (instead of replacing it with a phase annulus), and a large aperture can be used to obtain a shallow depth of focus in the specimen (whereas phase contrast suffers from a deep depth of focus which may superimpose different structures). Although each DIC objective normally has its own Wollaston prism installed, two or more condenser-mounted Wollaston prisms may be available. Thus, in a manner similar to phase-contrast microscopy, it may be necessary to change the lower Wollaston prism when the objective magnification is changed. It is useful to keep a chart next to the LM showing which lower Wollaston prism should be used for each objective.

On disassembling a DIC system, record the position of each component in great detail, because they can often be reassembled in a myriad of incorrect ways. The orientation of a Wollaston prism can be seen by examining it between cross polarizers. Because of their relatively high cost and popularity with other researchers, surplus DIC components may be difficult to acquire; however, in the surplus department, you may be fortunate enough to find a late 1960s DIC LM with large Wollaston prisms mounted in an analyzer slot far above the objective. These are a gold mine of high-quality components for polarized light microscopy and interference experiments, and they are easier to automate then a modern system with miniature Wollaston prisms.

1.7   PC Interfacing

The main types of interface likely to be found on a CAM are the general purpose interface bus (GPIB; HP-IB or IEEE 488) and the RS-232. When seeking solutions to problems with the GPIB it is generally best to look for the answer in the small print of operating manuals, while for RS-232 problems the fastest solution may be to use a breakout box and solder a new cable, as described below.

1.7.1   GPIB (HP-IB or IEEE 488)

General purpose interface bus hardware consists of sturdy 24-pin connectors with knurled knobs for tightening by hand. The screwdriver slot should only be used for loosening the knob. GPIB cables tend to be expensive and are much more difficult to fabricate by oneself than are RS-232 cables. There are 16 data lines accessed by OUTPUT and ENTER statements in software, plus 3 handshake control lines and 13 management lines accessed by SEND statements. With up to 15 devices connected on the bus, one of them must be designated as the active controller. The peripherals may be listeners, talkers, or both. Their addresses are set from switches, which usually are readily accessible on the rear of the instrument but sometimes are hidden internally. It is useful to advertise the current address of a peripheral using an up-front label, together with the assigned name of the I/O pathway used in software. This, of course, greatly facilitates substitutions, because the numerical address is only assigned once at the top of each program or subprogram. In some cases, the assignation can be made programmatically. A group of listeners can be assigned a common I/O, such as:

```
ASSIGN @Eavesdroppers TO 701, 702, 703
```

then all may be sent the same data with an OUTPUT statement, with data being accepted by different listeners at differing rates. With an ENTER statement, only the first device on the list becomes the talker.

For maximum reliability, all devices connected to the GPIB should be turned on, but disconnecting the cables of unused peripherals can become a chore. So, ignoring the aspirations of those who developed the GPIB, it may be convenient to use GPIB manual switch boxes to bring clusters of related peripherals on-line together.

Keeping an old PC as a slave on the GPIB may be very useful for low-level control of temperature and pH, but there can only be one active controller on the bus. PASS CONTROL can be used in software to switch from one controller to another. Sometimes the GPIB in a controller has a switch that makes it assume either active or non-active status at power-up.

An old controller also may be kept as a standby, isolated from the GPIB by a manual switch. Some computer-assisted microscopy experiments take hours to run and may use expensive reagents, so that a disaster near the end of a day's work is a worrying possibility. Lamp bulbs can burn out, arc lamps can lose their arcs after momentary power losses, and there are a dozen ways to alter the optical conditions of the LM with an inadvertent knock. While one strives to anticipate every contingency with ON ERROR statements and ON KEY backward looping, new crises appear at regular intervals. At such a time, it can be wonderfully soothing to the microscopist to put the active controller off-line, switch in the backup controller, write a few new lines of software to fix the problem, and then recover — which is why keeping all the standardization conditions in a COM block that can be updated from the keyboard is a smart way to run a CAM. In a windows environment, the same objective may be achieved by opening a new window, provided that I/O operations are not locked out.

Some of the peripheral devices used in a CAM may be operable in local mode from their own front panels. Control of each or all of them may be assumed from software with the REMOTE statement and released by a LOCAL statement. An obvious example is the scanning stage joystick, which must be in local mode to allow the operator to locate a specimen but should be inactivated by a REMOTE statement to prevent accidental operation later in the program. Quite often a device can be called back to local operation with a front-panel local-remote toggle switch, but not if it has been de-activated from software with a LOCAL LOCKOUT statement. The LOCAL LOCKOUT is a great aid in protecting the system from the absent-minded, especially for mechanical devices not protected by limit switches and for programmable power supplies. Even with local lockout in software, it is still wise to scatter slip-clutches and fuses through the hardware, knowing that often the greatest risk to a system comes from its own inventor and programmer.

The GPIB may be operated with various types of maskable interrupts (temporarily ignorable) and non-maskable interrupts (top priority emergencies) using the service request (SRQ) line (pin 10) of the GPIB. Interrupts enable the active controller to respond to situations arising peripherally. In addition, the active controller can solicit a status byte response from all peripheral devices with a parallel poll (PPOLL) statement and, if necessary, obtain further information from an individual device by means of a serial poll (SPOLL) statement. Servicing the device or fixing its problem may involve interrogating or altering the status of registers in the device, which may be done by extended addressing, adding the register address after the device address.

When all else fails, the first thing to look for is a panic button, such as the CLEAR I/O KEY of a Hewlett-Packard workstation or ALT F5 when using Windows. This may clear the problem, enabling the CONTINUE (ALT F2 from Windows) to carry on with the program. If this fails, the next thing is to type a CLEAR statement from the keyboard, addressing it to the device most likely to be causing the problem. As a penultimate resort, issuing an ABORT statement from the keyboard should terminate all bus activity and reset all devices to their power-on state. The likelihood of losing data already collected by the peripheral device increases with each of these steps. Situations such as this should not happen during normal operations using the GPIB, but they are quite likely to occur with a suspect new device. GPIB hardware is very sturdy, but not indestructible. With more than three connectors on top of each other, the leverage from someone tripping on the GPIB cable can cause subtle irregularities of performance behind the backplate of the device. One or more devices not powered up on the bus, or two devices with the same address, can cause erratic problems which are difficult to duplicate and fix. The final resort in these cases may be to turn off everything on the bus (including the controller), wait a bit, and then turn everything on again, concluding with the active controller which will need to test the GPIB at power-up.

1.7.2   RS-232

The serial interface known as the Recommended Standard 232, or RS-232, has acquired many nominal prefixes over the years (such as EIA for Electronic Industries Association, ANSI for American National Standards Institute, and TIA for Telecommunications Industry Association) but, like a new box for detergent or breakfast cereal, not much has changed on the inside. Electrical and mechanical interface requirements for the RS-232 are specified, but the lines to be used and how they are to be used are flexible (unlike the GPIB which should always work the same way). Different manufacturers use different patterns and rates of asynchronous data transfer and different methods of handshaking. Unlike using a GPIB device, a soldering iron generally is needed to connect a peripheral device to an unrelated PC via a serial port.

Step 1 is to dismantle the connectors at each end of a cable and see which pins are wired up. Usually, less than half the pins in a 25-pin connector are doing anything. The common ones are

(1) Safety ground or cable shielding, connected at one end

(2) Transmitted data

(3) Received data

(4) Request to send

(5) Clear to send

(6) Data set ready

(7) Signal ground, relative to 2 and 3, connected at both ends

(8) Data carrier detect

(20) Data terminal ready

The pins are numbered with microscopic numbers embossed in the plastic moulding that holds the connector pins. One of the useful features of a LM eyepiece is that, if taken from its tube and inverted, it becomes a high-quality magnifier. Thus, when held close to the eye, it is an ideal way to check the pin numbers. Always check the pin numbers before you touch anything with a soldering iron, because the mirror images of male and female connectors can become rather confusing.

If data are transmitted on line 2 and received on line 3, then lines 2 and 3 must cross over somewhere between the PC and the peripheral device. Checking this may be as simple as following the red and black wires from one end of a cable to another, but if the route passes through switch boxes or extension cable connectors, then it will have to be checked with an ohmmeter (after disconnecting all connections to computers and peripherals).

The idle state of the receive data line is usually high, equivalent to logical state 1. Thus, when data are transmitted, the first bit is a start bit with logic 0. Without the start bit, if the first data bit happened to be a 1, it would be missed. After the start bit may come 5, 6, 7, or 8 data bits, as agreed upon by both the PC and the peripheral device. Since the serial interface of the PC is programmable, whereas that of the peripheral is likely to be set by switches (if it is not altogether fixed), it is essential to find out how to program the serial port of the PC. And, while finding how to program the number of data bits, you also will see how to program the Baud rate, which must also be agreed upon by both the PC and peripheral device so that their oscillators are running at the same rate to catch each bit as it arrives. The most useful rates are likely to be 300 Baud (when communicating with an antique device), 1200 Baud (often a good compromise between speed and reliability), and 9600 Baud (top speed on a CAM, except for image analysis).

Sometimes a parity bit is sent after the data bits to check that the data bits were properly received. If this is being done by odd parity, then the sum of all the data bits plus the parity bit will be odd. For example, if the sum of the data bits is even, then the parity bit will be set to 1. This enables the receiving device to check that a single error has not occurred in transmission of the bit but, in the unlikely event of two or more errors per byte, they might all escape detection, depending on how they add up. Even parity works in a similar manner, but with an even sum for the data bits plus the parity bit. Parity can be programmed from the PC, turning parity checking on or off or setting parity to odd or even. Either 1, 1.5, or 2 stop bits may conclude the byte, allowing the receiver to do some housework before the arrival of the next byte. The chip that handles all these functions is the UART (universal asynchronous receiver-transmitter).

So far so good. We have checked that each data transmit line connects to a data receive line and that the PC is handling the correct number of data bits, with or without odd or even parity checking, as well as the correct number of stop bits. To do this, it was necessary to know how to program the serial port of the PC and how the serial port of the peripheral device was configured. If, by chance, both the PC and the peripheral happen to be using the same or compatible forms of handshaking, then communication now has been established, with the magic message "hello" flashing between PC and peripheral device.

If test messages do not get through, the problem is likely to be with the handshaking lines. At best, the logic may be obvious. Suppose that the sender may not transmit until "data set ready" (line 6) and/or "clear to send" (line 5) are high, while the receiver may not enter data until "data set ready" (line 6) and "data carrier detect" (line 8) are active. In which case, just as transmit data and receive data lines had to be crossed over, some of the protocol lines will have to be crossed. To give a specific example, here is the interchange required to operate an IBM-compatible stepper motor drive from a Hewlett-Packard (HP) workstation:

```
HP frame ground 1 → 1 device frame ground
HP transmitted data 2 → 3 device received data
HP received data 3 → 2 device transmitted data
HP clear to send 5 → 5 device clear to send
HP data set ready 6 → 5 device clear to send
HP signal ground 7 → 7 device signal ground
HP data carrier detect 8 → 5 device clear to send
HP data terminal ready 20 → 20 device data
                                   terminal ready
```

The HP serial port at Sc (select code) was programmed in HP BASIC:

```
CONTROL Sc,0; 1 ! put 1 in register 0 to reset port
CONTROL Sc,3; 9600 ! Baud rate for register 3
CONTROL Sc,4; 3 ! bit pattern for register 4, asking
                ! for 8 bits, no parity, 1 start, 1 stop
```

Working from an IBM environment and operating a different sort of device from the PC keyboard, one might program the serial port using:

```
MODE COM2:9600,e,7,1
MODE LPT1:=COM2:
```

to get 9600 Baud, at even parity, with 7 bits and 1 stop bit, sending something to COM port 2 as if it was line printer 1. In which case, IBM data terminal ready line 20 could be used as a high for device lines that were checked. It is hoped that this gives some idea of the sort of things that need to be done if everything involved is fully documented. Users of QuickBASIC will find the first few chapters of Nickalls and Ramasubramanian (1995) most instructive.

At worst, however, if handshaking lines were soldered in a granny knot somewhere deep inside an ancient cable, or the laboratory rat nibbled the PC handbook, it may be time for the breakout box. This handy device, which can be homemade if you wish, gets inserted in-line to make all the lines accessible as terminal posts, indicating their high or low states with light-emitting diodes (LEDs). Working at 300 Baud for both talker and listener (if possible), get the talker in an endless loop trying to send. Find which of the talker's handshaking lines needs to be made high before it will send blindly and repetitively along the transmit data line, whose LED will show very rapid flashes. With the listener in an endless loop trying to receive, find which of the listener's lines need to be made high before it will start accepting the data and flash it on screen. Initially, lines can be made high from any line that stays high all the time. Once communication is established, use the talker's intermittent high lines for the listener's handshaking input, and vice versa. Elevate the Baud rate to the desired rate on both the PC and the peripheral device, then slow down the listener's program with a WAIT or do loop, to check that the handshaking causes the talker to wait for the listener.

While delving into the technical literature for the PC, look for any error trap routines that can be incorporated into software as a debugging tool. The breakout box itself can cause problems by exposing the data lines to interference, which may appear as parity errors if parity checking is enabled. Framing errors occur if either the start bit or a stop bit have not been found properly. An overrun error, especially if it occurs when the listener's program has just been slowed down, may indicate that the handshaking is not yet correct, so that data have arrived at the listener in advance of an ENTER statement. The error code for break received may be very useful in future programming since it provides a convenient way of terminating data transmission from a peripheral to a PC, without the PC having to examine the content of every line of data looking for an embedded end of transmission message.

Having cracked a handshaking protocol, it is essential to make copious notes as remarks in the software. Then make two hard copies of the remarks, one to go in the laboratory diary and one to be folded up inside the housing of the newly soldered connector. Update your network diagram, giving the connector a unique number to match that on the diagram. Finally, write as much as you can on the outside of the connector housing. Chances are, you could save yourself a lot of time in the future.

Be as neat as possible when soldering up a connector and, in a microscopy laboratory, you may have the luxury of checking the work under a low-power dissecting LM. While soldering, support the connector in a blank connector of the opposite gender to act as a heat sink, use a small soldering iron, and cover the soldered connection with heat-shrink tubing. Tug on all soldered wires individually to check their integrity. Sometimes you may miss one, which then may wait until you undertake a critical or expensive experiment before it works loose. Shield cables in a grounded conduit around arc lamps and power supplies. And remember, when recycling used components from a basement or attic storage area, always avoid anything marked NFG (there is no polite interpretation for this widely used op-code).

RS-232 handshaking, however, is not always hard-wired and may be programmed using XON/XOFF. Buried in the appendices to the documentation for your PC is a list of ASCII characters. Decimal 17 (coded as DC1, for device control 1, with a hexadecimal value of 11h) is XON, while decimal 19 (DC2, hex 13h) is XOFF. The controller sends XOFF to stop a peripheral sending data, and XON to resume.

1.7.3   VXI

Reducing space and confusion is vital in the modern world of electronics, so a CAM is quite likely to involve instrument-on-a-card (IAC) systems. A card cage houses a set of cards, each of which replaces a separate device such as a multimeter, relay board, or analog-to-digital converter (ADC). The standard VME (Versa Module European) bus connectors and logical relationships developed for rapid data transfer in microcomputers are easily overwhelmed by the noise of a stack of multipurpose cards, and so a new standard called VXI (VMEbus eXtensions for Instrumentation) has been developed for IAC systems by leading manufacturers. VXI provides a superior interface between GPIB message-based devices and VME register-based devices, offering the high speed of low-level binary communication from a relatively simple high-level language. In slot 0 of a VXI mainframe is located a "commander" with the intelligence to drive the register-based "servant" cards in the other slots, as well as being the resource manager for VXI and GPIB. Thus, for operating the CAM, the subprograms may use the standard commands for programmable instrumentation (SCPI) to get very rapid responses from the servant cards. SCPI is a friendly language to incorporate into your favorite high-level language, as shown by a couple of BASIC examples:

```
OUTPUT @Multi;"MEAS:VOLT:DC?" ! go measure volts DC
ENTER @Multi;Answer ! and enter the answer

OUTPUT @Relay;"CLOS? (@103)" ! is relay 3 in slot 1
    closed?
ENTER @Relay;A$ ! 0 or 1 says false or true
IF A$ = "1" THEN OUTPUT @Relay;"OPEN (@103)" ! if
    shut, open it
```

As the older rack-mounted or stand-alone devices now clustering round most CAMs go up in smoke or refuse to awaken, they may be replaced by IAC systems.

1.8   Software

Operating a CAM may include getting a surplus stepper motor to operate at the end of a spare printer cable or putting a very expensive VXI card cage through its paces. Despite the four orders of magnitude difference in capital outlay, both activities may require about equal amounts of clear thinking and software, which brings us to the advisability of having an overview of CAM programming. With a little foresight, it is possible to write general purpose software to collect and analyze data from almost any conceivable experiment involving a CAM, thus minimizing delays between experiments, providing the maximum return on any software investment, and helping to clarify the logical design of experiments. Adopting a top-down approach to programming, the underlying similarities of a wide variety of experimental situations should be identified.

1.8.1   Graphics

Consider the output of a microscope spectrophotometer. A good way to start examining a spectrum, or any signal for that matter, is to plot it, and this requires graphics software. Thus, for a variety of different CAM applications, the graphics software is likely to be at the junction between the initial input and the later processing of data

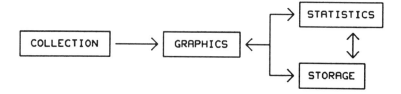

FIGURE 1.11
The importance of graphics in linking data collection to statistics and storage.

(Figure 1.11). It is important that this junction should never become an obstacle to innovation, and so our first consideration should be to anticipate the variety of forms in which incoming data may appear. A standard data format is required for all types of data to be graphed, but this must have maximum flexibility and also be convenient for data storage.

Looking forward to the time when there may be tens or hundreds of thousands of spectra stored together, they will be of little use unless they carry with them the coded information of how they were obtained and what they describe. For example, a *Fig* file might contain numerical data and have a matching *Leg* file containing text strings as legends. The *Leg* file may be given a fixed size and format to facilitate rapid searching, and it may contain a standardized string describing the project, experiment, sample, and replicate. This simplifies storage, retrieval, sorting, and editing of data once they accumulate. The variable dimensions of the *Fig* file might be coded in the *Leg* file (to simplify dimensioning matrices for the retrieval of the *Fig* file), together with information on how the spectra in the *Fig* file were collected.

For example, a spectrum in a file called *Fig1* and described in *Leg1* might have 21 measurements from 400 to 600 nm in increments of 10 nm with a bandpass of 10 nm. Regular spectra like this are the main product of many CAM applications and are very useful since they may be compared statistically. For example, a set of spectra might be tested against another experimental variable by calculating a t-statistic for a linear regression at each wavelength. Computationally, however, the t-statistic merely becomes the subject of a new spectrum to be displayed by the graphics software and stored. But, the subject of the graphics plot may not always be this regular, and wavelengths are not the only domains scanned with a CAM.

1.8.2   Different Types of Scanning

Polarized light accessories for the CAM enable the measurement of birefringence in plant cell walls or fibrous proteins of animal tissues. Birefringence may be measured by ellipsometry (Chapter 7), using measurements at different angles around an optical axis, rather than at different wavelengths. The operations involved in angular scanning are very similar to those involved in spectral scanning with a scanning monochromator and may be controlled with the same software using a common data format. In other words, there is little difference operationally between incrementing a monochromator and taking a measurement, and incrementing a polarized light analyzer and taking a measurement. But problems may be anticipated if angles rather than wavelengths appear on the x-axis. For example, it might be necessary to scan $-30°$ to $+30°$ each side of a setting at $10°$, in which case the scan will be from $340°$, through $360°$, to $40°$. In other words, it should not be assumed automatically that the correct order of x-axis data always is an ascending series.

Spatial scanning (incrementing the position at which the measurement is made) is another possibility, as in the stereological assessment of a three-dimensional system from data collected along a scanning transect. A scanning stage, where each increment in position follows a programmed sequence, produces a series of measurements at regular intervals, but other possibilities may arise as we extend the working range of the microscope using fiberoptics (Chapter 11). Hypodermic probes provide a novel way of obtaining data from within a system, be it an animal or a core of mud. A passive scanning device pushed into the sample by hand (to avoid the cost of an actuator or robot) may be triggered either at set increments in position or in time, and the latter option produces a series of measurements that are not spaced at regular intervals in position. Thus, in order to be able to process irregular vectors, it is unwise to reduce the x-axis to its determinants (minimum, increment, and number of values), something that might otherwise be done to save on storage space for a simple spectrum.

1.8.3   Dimensions of Scanning

A complex experiment may generate a multidimensional data matrix for display by the graphics software. Although it is relatively simple to assemble such a data set into a commercial graphics package from measurements that have already been obtained, custom programming is necessary to write software that can collect such data automatically. For example, using a CAM for fluorometry, a monochromator in the excitation pathway might be scanned to generate a fluorescence excitation spectrum, while a monochromator in the emission pathway might be scanned to generate an excitation spectrum. This can be handled using the concept of primary, secondary, or multiple scanners, in which case a response surface could be generated using the emission monochromator as a primary scanner to collect emission spectra, while the emission monochromator could be incremented as a secondary scanner after every primary scanning spectrum.

As another example, a series of ellipsometry measurements could be made along a transect across a sample. The primary scanner might be a rotary analyzer, with a scanning stage as a secondary scanner. This is a radically different example compared to the previous one (using excitation and emission monochromators for fluorometry), but data collection and the data matrix are very similar from the viewpoint of top-down programming. Determining the order of precedence for scanners depends largely on the apparatus involved and requires an understanding of the speed and accuracy of each scanner and the magnitude and impact of errors that it may create.

In subprograms for operating scanning devices, the primary algorithm generally includes a "do and wait" operation (for a stepper motor) or a "do and look" operation (for a servo system). Forward thinking is desirable so that errors do not occur when software is run on a controller that runs at a clock speed different from the one on which the software was developed. For example, for a new controller running at a faster clock speed than an old one, the faster "do" operation will require a longer "wait" if the scanner operation is matched to the duration of "do and wait". From this perspective, however, a scanner such as a monochromator or rotary analyzer is no different from any other device such as a timer or controller for sample temperature, pH, osmotic pressure, etc. With a little foresight, this enables a high degree of flexibility to be programmed interactively.

By adopting the concept of various orders of scanning devices and by having an interactive control program, it is relatively simple to reduce a complex experiment to a multidimensional matrix collected systematically by a series of CAM scanners. The operations involved in collecting optical data along a transect through a system and finding the effect of temperature or pH on the fluorescence of the same sample do not differ in nature — only in the number of matrix dimensions added by extra scanning devices. In other words, the operator might set a monochromator as a primary scanner (to obtain a spectrum), a pH changer as the secondary scanner (to obtain a response surface of spectra at a series of specified pH values), and a temperature changer as the tertiary scanner (to obtain a three-dimensional matrix by remeasuring the response surface at a series of programmed temperatures).

As every experienced programmer knows, the resources needed to expand a spaghetti-type program increase exponentially, while a neatly ordered top-down program can be expanded relatively easily. Time spent making every subprogram interrogate its inputs for array dimensions and CAM logicality will be amply repaid once the system reaches any degree of size and complexity. The art of survival is not to get trapped, which is why the CAM should be treated as three separate units: the LM, the interface, and the PC. The working life of a research LM can be measured in terms of at least several decades, while that of the interface may be about one decade. Who wants to get lumbered with a PC that old?

Chapter

Illuminators

2.1   Introduction

This short chapter draws attention to the importance of the illuminator in a CAM. Chapter 1 explained briefly how Köhler illumination became the dominant method of LM illumination once small, powerful automobile lamps became available. For a household mains current, a lamp filament must be long enough to match the voltage drop along the filament, whereas a low voltage allows a short filament pattern which can be imaged in the specimen plane; however, other types of illuminators are extremely useful for a CAM, as considered below.

2.2   Arc Lamps

2.2.1   Historical

By the 1930s, arc lamps were in regular use for high-intensity lighting of large public areas. Some of the smaller arc lamps were adapted for projecting microscopes used in teaching or morphometry. Two carbon rods were brought together and a start-up voltage > 44 V DC caused ionization, so that the two rods could be drawn apart to sustain an arc at a lower voltage. As in all arc lamps, a sturdy ballast resistance is required because, once formed, the arc has a negative resistance and requires a choke. The positive electrode was located above the negative electrode to form a vertical arc, with most of the light radiating from a crater formed on the tip of the upper electrode. The center of the arc produced intense violet light, surrounded by concentric zones of yellowish red, then pale green. The emission spectrum, however, was not smooth because it was modified by the spectral lines of the plasma. Naked arcs hissed alarmingly and required constant adjustment, so that Pointolite illuminators had many advantages for the LM.

The Pointolite illuminator had tungsten electrodes mounted in a partially evacuated bulb. Current was passed through a tungsten filament, coiled at one end and carrying refractory metallic oxides at the other. This ionized the surrounding space, enabling an arc to be struck away from the filament toward a third, globular tungsten electrode. The Pointolite illuminator was superseded by mercury vapor lamps, also derived from public lighting technology. Thus, the early ones were very large. But size was a disadvantage in microscopy, where only a small part of a large arc could be used. Thus, we arrive at the currently dominant technology, the high-pressure, short-arc lamp. Although lasers are essential for scanning confocal microscopy, they are poorly suited for conventional microscopy, where the field of view is reduced to a contour map of interference effects.

2.2.2   Short-Arc Mercury Lamp

Short-arc mercury lamps (usually designated as HBO) are used when high luminance in the visible range is required or when intense UV light is required for fluorescence microscopy. The lamp is a quartz glass tube with a bulbous expansion where the arc is formed. The lamp is filled with a base gas and an accurately measured amount of mercury which is vaporized to attain a high pressure when the lamp is operational. For DC operation, the negative electrode is uppermost while the lower positive electrode is larger to dissipate heat. The lamp is clamped at its lower end only, with a flexible electrical connection to the upper electrode. Older lamps had a third start-up electrode in a side arm, which greatly increased the risk of breakage when lamps were changed. The power rating of lamps for the LM is usually 50 or 100 W.

HBO lamps have a short operating life (such as 100 hours for the Osram HBO 100W/1) relative to other LM illuminators. Factors that decrease the operating life are operation for short periods of time (<2 hours) and a residual AC ripple current in the DC power source. Before attempting any alterations of the lamp mounting angle, the manufacturer's specifications should be checked, because some lamps have very narrow tolerances (such as 15° or 20° for the Osram HBO 50).

When the lamp is turned on, an arc discharge occurs in the base gas, followed by vaporization of the measured amount of mercury in the lamp. The initially wide luminous arc becomes narrower as the mercury vapor pressure rises. The minimum operational warm-up time is 15 minutes. If the arc is lost because of a momentary disruption in the power supply, check the manufacturer's instructions before attempting to re-start the lamp. In older lamps, when they are hot, the resistance is low and may damage the start-up circuit. If in doubt, wait until the lamp is cool before re-starting it. Neither the heat nor ozone production of XBO lamps is excessive, and they are usually cooled by natural convection currents of air passing through louvers in the lamp housing. Natural convection currents may be enhanced with an extractor fan above the lamp, but cold air should not blow directly on the lamp tube.

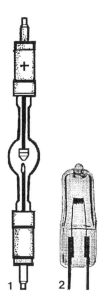

FIGURE 2.1
An XBO xenon arc (1) compared with a halogen tungsten-filament lamp (2).

2.2.3   Short-Arc Xenon Lamp

Usually designated as XBO, the short-arc xenon lamp has a quartz glass tube filled with pure xenon, the pressure of which increases when the lamp becomes operational (Figure 2.1, 1). The positive electrode is located on top and, for heat dissipation, is larger than the lower negative electrode. Reversed installation will destroy the lamp almost instantly. Operational life for XBO lamps is much longer than for HBO lamps (400 hours for the Osram XBO 75W/1), but they become blackened with age so that luminance may decrease. Blackening, mainly in the upper part of the tube, is increased by uneven heating, as may happen if the mounting angle is beyond the manufacturer's specifications. XBO lamps often are power-cooled, especially those for UV illumination which produce high levels of ozone. Thus, an extractor fan venting to the exterior is highly desirable. If not, it may be wise to locate an ozone test card in the operator's environment to evaluate the health risk (obtainable from Vistanomics; Glendale, CA). Because of the pressure within an XBO lamp, protection for the face and hands is required when lamps are changed. Usually, the lamp is installed with its plastic safety case remaining on until the end of the installation. The safety case should be kept on hand for removal and disposal of the old lamp. All used lamp bulbs should be disposed of properly as hazardous waste, following institutional rules, but this is especially important for XBO and HBO lamps which, potentially, may be either explosive or toxic.

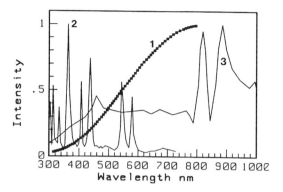

FIGURE 2.2

Manufacturers' emission spectra (relative radiant power) of three light sources: a 12-V, 100-W halogen filament lamp operated at 3400 K, 1; a short-arc mercury lamp, 2; and a short-arc xenon lamp, 3.

2.3   Tungsten-Halogen Filament Lamps

The advantage of filling a lamp bulb with a halogen, usually iodine, is to help recycle tungsten vapor from the filament (Figure 2.1, 2). When operated at a reduced voltage to dim the illuminator for low magnification, the recycling may be reduced because the quartz glass tube is not as hot as at full voltage. A stabilized DC power supply is needed for CAM operation, so it makes sense to operate the lamp almost at its maximum voltage, adjusting light intensity with neutral density filters in the light path.

2.4   Emission Spectra

Figure 2.2 shows the emission spectra of the illuminators considered above. The smooth spectrum for the 12-V, 100-W filament lamp (Figure 2.2, 1) is the most useful for spectrophotometry because there are no sharp emission peaks, as for the other two spectra. Intense peaks cause major problems when scanning measurements are made across the spectrum. If the photometer is set to a maximum reading on the most intense peak, then the light intensities in the valleys are so low that only a small fraction of the dynamic range of the photometer is used. Also, small mechanical errors by a monochromator in arriving at a wavelength will cause a major variation in light intensity if the wavelength is near an intense peak. However, the filament source (Figure 2.2, 1) has very little intensity at 400 nm, which may create problems for spectrophotometry at high magnification using a small photometric aperture.

Short-arc mercury lamps (Figure 2.2, 2) are widely used for fluorescence microscopy, especially if the highest peak at 365 nm can be exploited. The sharp peaks in the visible part of the spectrum make it difficult to use a short-arc mercury lamp as a source for conventional spectrophotometry, although many of these peaks are ideal for exciting fluorescence. One of the peaks, at 365 nm, is very close to the excitation maximum of collagen, thus providing strong excitation of connective tissue fluorescence (Chapter 8).

The spectrum for a short-arc xenon lamp (Figure 2.2, 3) is somewhat of a compromise between the previous two emission spectra. It has useful high intensities around 400 nm, a relatively flat plateau across most of the visible wavelengths, and some major emission peaks in the NIR that can be exploited or avoided, depending on the application. It is important, however, to remember that these spectra may be modified by heat absorbing or reflecting filters in front of the illuminator. The XBO spectrum is particularly suitable for measuring UV fluorescence excitation spectra (Chapter 8), where filament lamps are too weak and the HBO is too irregular.

2.5   Control of Illumination Intensity

Few experiments with a CAM require programming of the light intensity. It is relatively simple to control the current to a filament lamp, using a programmable power DC power supply (such as the HP 6642A); however, reduction of the current changes the shape of the emission spectrum so the current cannot be modified during spectrophotometry.

Programmable dimmer devices were available in the early days of video, when cameras required a relatively constant light intensity to maintain their color balance, despite changes in objective magnification. The principle of operation was to slide two graded neutral-density filters towards each other, thus decreasing the obvious difference in brightness from left to right and making the center of the field brighter than the edge.

If a polarizing CAM is available, illumination intensity may be controlled by programmable rotations of the analyzer relative to the polarizer, with the added advantage of using the extinction coefficient to program transmittance (Chapter 7). For critical applications, a triplet of polarizers may be needed (Piller, 1981). Another method is to use a solenoid-operated, stray-light filter changer from a grating mono-chromator (Chapter 5). The stray-light filters are replaced with neutral density filters or with different grades of wire-mesh grids if working immediately in front of a powerful arc lamp (after checking that grids do not result in uneven illumination across the field).

Chapter 3

Shutters and Apertures

3.1   Introduction

Shutters and apertures are used in a variety of CAM operations. A shutter is necessary to find the dark-field output of a PMT, and a field aperture is necessary to reduce unwanted glare in photometry. Seldom, though, are CAM apertures programmable, in the sense that their diameter can be programmed. At the start of an experiment, the operator usually mounts an appropriate fixed-size aperture in a solenoid-operated swing-in frame (which might equally well hold a solid, unperforated disk to become a shutter). The apertures are programmed to swing in or out of the optical axis at various points during the measuring protocol. Thus, a field aperture may be moved out of line any time the operator needs to see the whole field to locate a specimen and then swung back into line any time a measurement is made. Light passes through LM apertures at an angle (cone of illumination) rather than as parallel rays. This may cause a small systematic bias on transmittance measurements, depending on the level of transmittance (T), numerical aperture (NA), and photometric aperture radius relative to the thickness of the sample (Boguth and Piller, 1988). However, in most situations, this may be ignored (when $T < 0.1$, $NA < 0.45$, and photometric aperture radius < 0.2 sample thickness).

The actuators that move shutters and apertures range in speed from relatively slow solenoids acting on a lever, for a swing-in shutter, to the high-speed solenoids of camera-type shutters. Thus, the interaction between shutter speed and photometer speed may be quite complex. A fail-safe attitude can be adopted by the programmer, assuming that the shutter or aperture takes longer to shut than it actually does and by assuming that the photometer takes longer as well. Not much time will be wasted, because it is not often that a dark-field measurement is required, and a field aperture may stay in-line for a whole measuring session. But, speed interactions may become important if a specimen is to be protected from strong illumination right up until the

35

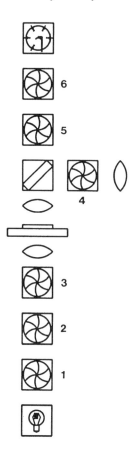

FIGURE 3.1

Shutter and aperture locations: illumination shutter, 1; field aperture for normal use, 2; photometric field aperture, 3; ocular shutter, 4; photometer shutter, 5; and photometric measuring aperture, 6.

point of illumination as, for example, when dealing with light-sensitive living cells or with rapidly quenched fluorescence. This creates some interesting challenges and opportunities, as considered in this chapter.

3.2   Shutter and Aperture Locations

Having several shutters in a CAM enables some useful diagnostic software tools to be written and may greatly enhance the sophistication of CAM applications. All possible locations of shutters and apertures are shown in Figure 3.1 where, as in the rest of the book, a symbol for a closed iris diaphragm is used for both shutters and apertures, regardless of whether or not they are manual or automated, or swing-in or segmented.

3.2.1   Illumination Shutter

This may be immediately in front of the illuminator (Figure 3.1, 1) or in front of the main axis of a mirror switching between two or more illuminators. When the illumination shutter is closed, the signal from the PMT is the dark-field current plus ambient light. Sometimes the ambient light can be quite high and quite noisy, especially if the room lights are on, the observer's ocular shutter is open, or a low-power objective with a long working distance is being used.

3.2.2   Photometric Field Aperture

In addition to the manual field aperture used to establish Köhler illumination (Figure 3.1, 2), a swing-in carrier for photometric apertures may be used as a supplement (Figure 3.1, 3). Its radius in the field of view is generally 2 or 3 times that of the small measuring aperture radius under the photometer head (unless, for some reason, a very large measuring aperture is required). When Köhler illumination is established for photometry, the photometric field aperture is the one that should be brought into sharp focus, not the normal manual aperture for ordinary viewing. The photometric field aperture reduces unwanted stray light from the field around the measured area. Examination of a well-engineered, swing-in field aperture should reveal some centering adjustments, as well as an adjustable mechanical stop that limits the maximum throw of the solenoid. In other words, unlike shutters that must merely flap across the optical axis, the photometric field aperture must be a precise device. If, after it has been swung out from the position at which it was centered, it swings back in to a different position, the amount of stray light may change. If the peripheral field excluded by the photometric field aperture is bright or asymmetrical, the error will be quite noticeable. At worst, the field aperture might overlap the photometric aperture. This can happen if the locking nut for the mechanical stop is loose. Thus, every time the aperture is swung in, it hits the stop screw and may turn the screw a small amount, so that the field aperture gradually progresses across the field of view. Thus, a whole experiment may be ruined.

3.2.3   Ocular Shutter

There are many situations when almost all the ambient light reaching the photometer can be eliminated with a shutter beneath the oculars (Figure 3.1, 4). Light may enter the oculars from an overhead room light to be directed downwards to the specimen. But, at each of the slightly reflective surfaces formed by the numerous lenses on the way down to the specimen, a small amount of light may reflect back upwards. Some of this reflected light passes back through the beam splitter to the ocular, but the remainder may reach the photometer. If the amount of light entering through the oculars is far greater than that from the specimen, the effect may become a major source of error. For critical experiments, the CAM must be used in a dark room. The ocular shutter is generally a manual device used at the discretion of the operator. For example, if it is opened at an inappropriate moment during a measuring protocol when the PMT is at high voltage, the PMT may be saturated. The problem can be avoided by using a direct pathway from the specimen to the photometer (not splitting the light to both the photometer and ocular). This is also required when light levels are extremely low, but now the operator cannot check the condition or centering of the target.

3.2.4   Photometer Shutter

The photometer shutter is located somewhere under the photometer head (Figure 3.1, 5). When it is shut, it gives the true dark-field output of the photometer. The magnitude and source of ambient light may be assessed by watching the photometer output as the illumination, photometer, and ocular shutters are closed. This is a useful test to undertake before the CAM is finally standardized for making measurements and tells the operator whether or not it is necessary to work in darkness or to close the ocular shutter during measurements. Often the results of this test are surprising. You might find that room lights have no effect, even though light levels are low because of high magnification. On the other hand, you might find that you have a sample chamber that draws in every photon of ambient light, so that the PMT responds when a bird flies past the laboratory window.

3.2.5   Photometric Measuring Aperture

The photometric measuring aperture is a precision device that limits the field from which light is measured by the photometer (Figure 3.1, 6). It is not often that the size of the photometric measuring aperture is changed during a series of measurements, so it is usually changed manually, not by the PC. It is good practice to insist on the operator declaring the size of the measuring aperture and the magnification of the objective. This information may be stored and used to recreate the same measuring conditions at a later date. In combination with wavelength (which gives the likely intensity of illumination), photometric measuring aperture size and objective magnification may be used in software to optimize photometer standardization. For example, with a small photometric measuring aperture, a ×100 objective, and violet light from a halogen lamp, it is a waste of time to start looking at low-voltage, low-gain outputs when setting up the PMT. Conversely, with situations leading to high light intensities, avoiding action can be taken in advance to prevent PMT saturation.

3.2.6   Secondary Illuminator Shutter

It may be necessary for spectrofluorometry to have a standard halogen lamp operated at 12 V to provide a spectrum for which the relative spectral intensities are known.

It is important to stabilize the lamp, so it must be left on all the time. When the standard is required, the illumination shutter (which now will be in front of an arc lamp) is closed, and the secondary illuminator shutter is opened. It is convenient, therefore, to have the secondary illuminator shutter under computer control so that the CAM can be re-standardized automatically.

3.2.7   Protective Shutter

The protective shutter may be an illuminator shutter described above or a separate entity, and it is used to protect the specimen from intense illumination. It is opened immediately before a measurement and closed immediately afterwards but stays open continuously for standardization (before the specimen is installed). Protective shutter operations may conflict with normal operations, so that having a separate protective shutter designation allows the logic of shutter control to be simplified, as described later. The protective shutter is given the highest logical priority and is operated after other shutters with a lower priority. The noise of a protective shutter constantly opening and shutting may become annoying, and it makes it impossible to check on the state of the specimen between measurements. Thus, one may choose not to install or designate a protective shutter at the start of an experiment, only to become aware that the specimen is being altered in some way by continuous, strong illumination.

3.3   Shutter-Photometer Interactions

A CAM may be programmed to make many different types of measurements, but it is difficult to control several different types of operations simultaneously with only a single controller. Essentially, the active controller steps through all its tasks in sequence and, unless interrupted, must finish one before moving on to the next. Thus, it is difficult to make an uninterrupted continuous series of measurements from a PMT at the same time that other devices such as shutters and apertures are being operated. One way around the problem is to use some type of parallel processor, such as a multiprogrammer with the capability for operating different cards simultaneously. Once instructed and triggered, the cards can perform their functions simultaneously and asynchronously, then wait to have their data uploaded sequentially by the active controller. Some degree of memory buffering is required somewhere in the system, in the ADC itself, in the card-cage controller, or as a separate card in the cage. The availability of a multiprogrammer, even if only temporary, provides a unique insight into the operation of a CAM, allowing programmed methodological studies to be undertaken while the PMT is working at its ordinary, assigned tasks. This allows the programmer to answer questions such as how long it takes for a PMT to equilibrate after a shutter has been opened. Experiments from Swatland (1993) are

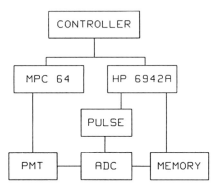

FIGURE 3.2

Block diagram of hardware for examining PMT response in parallel to normal CAM shutter operations, showing a typical CAM interface (MPC 64) between the controller and PMT, with a parallel circuit created by a multiprogrammer (HP 6942A) containing a memory board buffering the output of an ADC triggered by a pulse-train generator.

reported below to give some idea of the operations that are possible, leaving the reader to adapt specific components near at hand for a similar purpose if so required.

3.3.1   Hardware

An overview of the system is shown in Figure 3.2. There are many ways in which this layout could have been achieved, but in this specific example the main controller was an HP BASIC workstation linked via the GPIB to a Zeiss MPC 64 microscope interface to provide all the normal functions of a CAM, including the operation of the PMT. In addition to these ordinary CAM components, however, the PMT output also was connected directly to an ADC (HP 69751A) in a multiprogrammer (HP 6942A) to provide a dual measuring system. The ADC was triggered from a programmable pulse train generator (HP 69735A; period accuracy ±0.01%) used as the clock for time-based measurements. Data were stored on a 4-K memory card (HP 69790B) using first-in, first-out (FIFO) pointers to upload to the controller at appropriate times. Multiprogrammer cards were programmed by the armed-card interrupt method, the concept of which is that a card may be given an instruction (so that it is armed), but it will not perform the instruction until triggered in some way, after which it places a service request on the GPIB to upload its data to the active controller. The HP Multiprogrammer User's Guide 06942-90013 may be consulted for further details, although it might be wiser to look at newer instruction sets, such as those for the VXI.

A Zeiss Universal microscope with a range of accessories was used as basic equipment. The PMT was a side-window Hamamatsu HTV R 928 with S-20 characteristics. It was mounted on a type 03 photometer head (Zeiss 477310) with invariable measuring apertures (Zeiss 477322). A grating monochromator (Zeiss

474345) was used with bandpass of 10 nm. A 12-V, 100-W halogen source (Zeiss 487332) with a stabilized transformer was used for shutter experiments and for standardizing the photometer for measurements of relative fluorescence intensity (Zeiss, 1980). A 100-W mercury arc (Zeiss 482590) was used for fluorescence excitation.

3.3.2   Software

In the following BASIC subprogram, the array Mpc(*) contains the operating variables for the PMT, as discussed in Chapter 4. SUB Memin uses SUB Shutter on line 540 to flap a shutter out and back into the light path as fast as it can. The commands for opening and closing a shutter are very simple; for a Zeiss fast shutter (467225) they are

```
OUTPUT @Mpc USING "#,A";"o" ! open fast shutter
OUTPUT @Mpc USING "#,A";"n" ! shut fast shutter
```

This creates a pulse of light to the PMT and is the sort of thing that might be done to get a fluorescence measurement from a specimen whose fluorescence is rapidly quenched by exposure to the light that excites the fluorescence. This keeps the active controller completely busy dealing with the shutter and is the equivalent of a normal CAM operation, but some while in advance of this the memory cards were linked to their respective ADCs and formatted (line 320). And immediately before SUB Shutter was called, the pulse train acting as a trigger for the ADC was started (line 520). If a really fast shutter is in use, a short delay is needed (line 530) to make sure all the action is caught in the graphics frame. As the shutter is being flapped by the active controller, previously armed cards in the multiprogrammer collect data as they are triggered. On line 730, SUB Memin unloads the data from one of the cards using FIFO pointers, creates an x-axis (line 970), and uses it to find the area under the curve for the pulse of light (line 1000).

```
10    Memin:SUB Memin(Mpc(*),Area,Dark_flag)
20    !
30    ! [1] Mpc(*) holds operating variables, see
      Chapter 4
40    ! [2] Area takes data back into main program
50    ! [3] Dark-flag identifies the dark-field shutter
60    ! [4] 2 cards are triggered, only data from 1
      is used
70    ! [5] ADC end of cycle connects to data
      available for memory
80    ! [6] ADC and memory cards must share a common
      ground
90    !
100   N=100 ! fixed number of data wanted for simple
      example
```

```
110   Le=Mpc(25)+.1! length of recording window
120   ALLOCATE A(N-1) ! only going to use these data
      from 1 card
130   ALLOCATE B(N-1) ! could be used to study other
      CAM device
140   ALLOCATE Xdat(N-1) ! x-axis data for SUB Area
150   !
160   ASSIGN @Multi TO 723 ! addresses and extended
      addresses
170   ASSIGN @Multi01 TO 72301
180   ASSIGN @Multi05 TO 72305
190   ASSIGN @Multi06 TO 72306
200   ASSIGN @Multi10 TO 72310
210   ASSIGN @Multi12 TO 72312
220   ASSIGN @Multi14 TO 72314 ! to read clock
230   !
240   ! how to format these memory cards
250   ! --------------------------------
260   ! set format (SF) sends:
270   ! [1] the address of the card
280   ! [2] the number of formatting parameters
290   ! [3] data type (1 = decimal 2's complement
      binary)
300   ! [4] LSB of input (normally .005 from this
      ADC)
310   ! [5] size of input (always 12 bit from this
      ADC)
320   OUTPUT @Multi;"SF,2,3,1,.005,12T" ! format mem
      in slot 2
330   OUTPUT @Multi;"SF,5,3,1,.005,12T" ! format other
      one
340   !
350   ! how to program memory cards in slots 2 and 5
360   ! --------------------------------------------
370   OUTPUT @Multi;"WF,2,1,3",N-1,"T" ! set FIFO &
      ref register
380   OUTPUT @Multi;"WF,5,1,6",N-1,"T" ! same for slot
      5
390   OUTPUT @Multi;"WF,3.1,0,3.2,0,3.3,0T" ! card 2,
      zero pointers
400   OUTPUT @Multi;"WF,6.1,0,6.2,0,6.3,0T" ! same for
      slot 6
410   !
420   ! how to arm the memory cards
430   ! ---------------------------
440   OUTPUT @Multi;"AC,3,6T" ! arm both, terminate
      with SRQ
450   !
460   ! how to set up pulse train period as a trigger
```

```
470   ! --------------------------------------------------
480   OUTPUT @Multi;"WF,11.1,0T" ! write first, using
      period * 1
490   Period$=VAL$(1000000*Le/N) ! Period$ = pulse
      train period
500   ! note how microsec period = 1E6 * window / no.
      in window
510   OUTPUT @Multi;"WF,11.2,"&Period$&"T" ! micro-
      seconds
520   OUTPUT @Multi;"WC,11",N,"T"! start train trigger
      with N pulses
530   IF Mpc(24)=2 THEN WAIT .05 ! wait if a really
      fast shutter
540   CALL Shutter(Mpc(*),7) ! flap whatever shutter
      was specified
550   !
560   ! wait for interrupt after it is all over
570   ! ----------------------------------------
580   Wait:ENTER @Multi10;U,V,W
590   IF W<2 THEN Wait ! wait for interrupt
600   !
610   OUTPUT @Multi;"WF,2.1,24T" ! lock-out further
      data, otherwise
620   OUTPUT @Multi;"WF,5.1,24T" ! stack pointers get
      messed up
630   ! note how 2.1 is address + subaddress
640   ! note code is 20 for lockout + 4 for FIFO
650   CALL Cheker(U,V,W,X,Y,Z) ! what's up on the GPIB?
660   !
670   ! read from cards that interrupted
680   ! -------------------------------
690   ENTER @Multi12;I,J
700   OUTPUT @Multi;"DC,3T"! disarm card, so we can
      do something
710   OUTPUT @Multi;"DC,6T"! disarm other card as well
720   OUTPUT @Multi;"MI,2",N,"T" ! MI says card 2
      memory input of N
730   ENTER @Multi05;A(*) ! here are data from first
      card
740   OUTPUT @Multi;"MI,5,",N,"T" ! another MI link
      and number
750   ENTER @Multi05;B(*) ! here are data from second
      card
760   GOTO Meminend ! jump over diagnostics
770   !
780   Readmem: ! de-bugger used to follow stack
      pointers
790   OUTPUT @Multi;"RV,3.1,3.2,3.3T" ! RV says read
      values
```

```
800   PRINT "----rv of mem slot 3-----"
810   GOSUB Flash
820   OUTPUT @Multi;"RV,6.1,6.2,6.3T"
830   PRINT "----rv of mem slot 6-----"
840   GOSUB Flash
850   PRINT " "
860   RETURN
870   !
880   Flash:ENTER @Multi06;R1,R2,R3 ! get values to
      show us
890   PRINT "differential counter at ";R1
900   PRINT "write pointer at ";R2
910   PRINT "read pointer at ";R3
920   PRINT "--------------------------"
930   RETURN
940   !
950   Meminend:MAT A= ABS(A) ! no neg numbers if ADC
      wires crossed
960   CALL Screen(A(*),Le,N-1,Test) ! plot data on
      screen
970   FOR I=0 TO N-1 ! shows how x-axis is made from
      pulse
980   Xdat(I)=I*(Le/N) ! pulse period and is used
990   NEXT I
1000  CALL Area(Xdat(*),A(*),Area) ! to find area
      under curve
1010  !
1020  SUBEND !-----------------------------------
```

3.3.3   PMT Response

A mechanical camera shutter (five-segment iris) mounted in the illumination pathway was used to test the system (simply by changing Dark_flag to a code for an electronic camera shutter). Figure 3.3 shows the response of the PMT to white light with a series of exposures of increasing duration (transmittance = 1, with continuous light). Using the ADC in the multiprogrammer it was possible to follow the PMT output and to find that a delay of about 0.1 sec was required before taking a measurement through the normal circuit (Figure 3.4).

Figure 3.5 shows the effect of shutter speed on the area under the peak and the peak width at half its height. Area and peak width both were correlated with shutter speed, $r = 1$. Thus, for situations where photobleaching or deleterious sample responses may occur (as with living tissues), either area or peak width may be used to make rapid measurements without waiting for the PMT to equilibrate; however, it cannot be assumed automatically that relationships such as those shown in Figure 3.5 will occur at all operating conditions. Do they change with wavelength, high voltage, or gain?

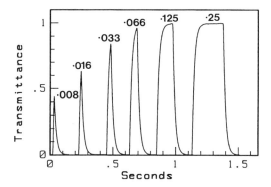

FIGURE 3.3
PMT response to illumination modulated by a mechanical camera shutter (exposure times in seconds).

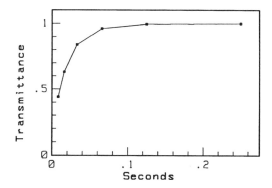

FIGURE 3.4
Effect of exposure time on maximum PMT response.

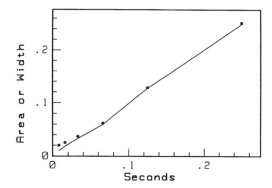

FIGURE 3.5
Effect of exposure time on the area under the peak (line) and the width of the peak (solid squares). Peak width was measured at a level halfway up the peak.

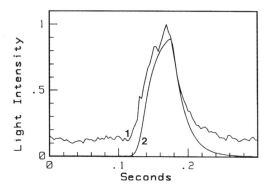

FIGURE 3.6
Response to opening a shutter with the PMT at high voltage and high gain (1) and at low voltage and low gain (2).

Figure 3.6 shows the PMT response at extremes of high and low light intensity, comparing white light through a large photometer aperture (halogen source, ×10 objective, 0.32-mm photometer aperture, 575 V, ×10 gain) with monochromatic light through a small aperture (450 nm, 10-nm bandpass, 35% grating efficiency, 0.08 aperture, 1107 V, ×1000 gain). With a low light intensity, the PMT current was noisier and had a high dark-field current, but the overall shape of the peak was similar. After subtracting the minimum dark-field current, the areas under the peaks differed by only 1.25% and peak width was identical. Thus, over the working range of this particular type of S-20 PMT, one would not expect serious errors using this method to minimize exposure of the sample to light.

A check was made to ensure that changes in wavelength did not invalidate using areas and peak widths for spectrophotometry (rather than using pulses of white light, as described in the previous paragraph). This is a difficult measurement to make directly because a change in wavelength causes many other factors to change. The wavelength-related factors that are involved include: (1) the voltage-dependent emission spectrum of the halogen illuminator, (2) the summation of transmittances of all the glass lenses between the illuminator and the PMT, (3) the grating effi-ciency of the monochromator, and (4) the spectral response of the PMT. However, when all these factors are combined to produce an overall spectral response, each position on the ascending slope must have a corresponding position on the descend-ing slope. In other words, taking a low wavelength and a high wavelength that both produce an identical steady current in the PMT, what happens when the PMT is exposed to only a short pulse of light through a shutter? When tested across the whole of the working range of the CAM, the response to a 0.1-sec pulse at 394 nm was identical to that at 722.5 nm (using a high-performance, solenoid-operated iris shutter to produce a response intermediate between those shown for 0.066 and 0.125 sec in Figure 3.3). Looking near the center of the working range, the voltage of the illuminator was reduced until 600 nm gave the same PMT steady current as that for the first pair of wavelengths. A pulse of light at 548 nm, the matching wavelength to 600 nm, gave an identical response. Thus, the responses at 394, 548, 600, and

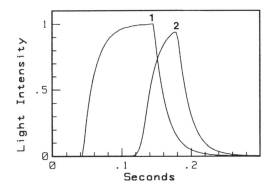

FIGURE 3.7
A five-segment iris shutter (1) compared with a solid shutter (2), both operated with the same delay (0.05 sec) and open time (0.1 sec).

722.5 nm were identical, and it was concluded that wavelength-related sources of error were negligible over the normal working range when the PMT voltage and gain were constant.

3.3.4   Comparison of Two Shutters

Two types of solenoid shutters may be found on a CAM: a solid shutter that swings unidirectionally across the optical axis and a segmented iris shutter that closes in a centripetal pattern. The solid shutter generally is less expensive and has a lower performance than the iris, but only the solid shutter may be used for holding apertures to define the field of view. Solid and iris shutters (Zeiss 467226 and 467225, respectively) are compared in Figure 3.7.

Both shutters were programmed with a 0.05-sec delay before opening and to stay open for 0.1 sec, all within a measuring window of 0.3 sec. The PMT response to the high-performance iris shutter (Figure 3.7, 1) started with almost a linear rise at 0.042 sec as the five segments of the iris withdrew from the optical axis. Since the 0.05-sec delay was started after the multiprogrammer was configured, this showed the time required for this operation (about 0.008 sec). The linear rise gave way to a curve created by the saturation characteristics of the PMT. In contrast, the solid shutter (Figure 3.7, 2) had a long delay before the curved edge of the shutter started to expose a crescent-shaped opening between the edge of the shutter and the inner curve of the aperture. Thus, the signal started like a sine wave. The solid shutter did not expose the PMT long enough to reach its maximum.

3.3.5   Minimum Exposure Protocol

As discussed later in Chapter 4, a typical algorithm for setting a photometer is based on a number of programmable parameters that may be user-specified or fixed in the

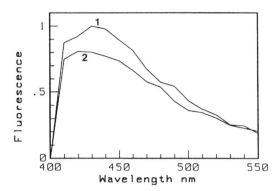

FIGURE 3.8

Fluorescence emission spectra of a bovine collagen fiber measured by the minimum exposure method (1) and remeasured after exposure to UV light for 60 sec (2).

software. For a PMT, the following might be used. First, the fraction of the dynamic range to be used must be established. Use of the full range gives the most gray levels, but risks saturation of the PMT. For example, a fraction of 0.9 might be used for accurate studies on a known sample, while 0.5 would be safer for a first examination of an unknown sample. Knowing the fraction of the dynamic range to be used, the gain and high-voltage of the PMT then may be selected, but this will require a window of acceptance. In other words, the algorithm may be terminated with a PMT output that is within a certain distance of the maximum PMT output. Signal noise may be canceled by using a certain number of replicated ADC cycles, and the response time of the PMT may be taken into account by only accepting signals after successive new measurements cause less than a specified change in the signal average. In other words, the requirements for precise, accurate measurements call for relatively long periods of sample illumination.

For rapid measurements with a minimum exposure of the sample, the standard method is to open a shutter with a single instruction, followed immediately by a double instruction to cycle the ADC and to close the shutter. It is difficult to correct for noise at low light intensities if there is only one measurement from an undamped signal. A photodiode array spectrometer synchronized with a xenon flash unit is probably the fastest way to measure a photosensitive sample (Chapter 10), but it is still worthwhile enhancing the performance of a PMT-based CAM, which then may be used for all but the most demanding photosensitive samples.

Figure 3.8 shows two fluorescence emission spectra of a collagen fiber dissected from a bovine tendon and mounted in distilled water. The fiber was located using transmitted white light from a halogen source, then its fluorescence was measured with a type III RS epi-condenser (×25 neofluar, 365-nm excitation filter, 395-nm dichroic mirror, 2.5-mm photometer aperture). The photometer was standardized from a halogen source, using a small area of aluminum foil mounted next to the sample as a fluorescence blank (Chapter 8).

The first spectrum (Figure 3.8, 1) was created by finding the area under the photometer signal during an exposure of 0.1 sec for each wavelength so that, from 400 to 550 nm with 10-nm increments, the total exposure time was 1.6 sec. Immediately after obtaining spectrum A, the fiber was exposed to UV light for 60 sec (because native tendon does not quench rapidly) and then remeasured to obtain spectrum B (Figure 3.8, 2). The decline in fluorescence of spectrum B relative to A was caused by fluorescence quenching. Thus, had the sample been measured by the normal method requiring a relatively long period of exposure of at least 60 sec, the result would have been intermediate between A and B. The convergence of spectra A and B towards 550 nm might have resulted from the recovery of fluorescence during the second sequence of measurements when the sample was exposed for only 1.6 sec during a measuring time of 76.6 sec.

3.4   Shutter Software for Normal Operations

With a relatively simple CAM, it is possible to write shutter close and open instructions directly into the software for the measuring protocol, using direct register addresses. At this stage there may only be one simple shutter. But if CAM experiments prove successful, as it is hoped that they will, it may not be long before other shutters are added for various different operations. Thus, it is wise to gather all shutter operations into one sub-program, allocating I/O pathways in a location that makes them easy to read and simple to change. The final sophistication that becomes very powerful, once the effort has been made to establish the system, is not to instruct the SUB shutter to open or close a specific shutter, but rather to request one of a class of operations, as listed below. When the optical layout is planned at the start of operations, each type of shutter and aperture is identified interactively. This allows almost any CAM layout to be operated from one program, a program that will not let the operator do anything foolish, such as attempting to use a CAM layout with no dark-field shutter. It is important not to have a specific shutter or aperture at a specific location or with its I/O calls hidden in a rarely used subprogram. Sooner or later, it may open a shutter unexpectedly and saturate the PMT.

3.4.1   Power-Up and Power-Down

Open all shutters and apertures except: (1) the photometer shutter, to avoid exposing the PMT to intense light from a normal full-field illumination; and (2) the secondary illuminator shutter.

3.4.2   Measuring Position

The illuminator shutter should be open, the secondary illuminator shutter (if present) should be closed, and the photometric field aperture should be located in the optical pathway. A prompt may be needed for the operator to close the ocular shutter, if necessary. Finally, the photometer shutter is opened.

3.4.3   Viewing Position

First of all, the photometer shutter must be closed to protect the PMT. Then the illuminator shutter should be opened, the secondary illuminator shutter (if present) should be closed, and the photometric field aperture should be out of the field of view. The operator may need a prompt to open the ocular shutter if previously it was closed.

3.4.4   Visible Light Viewing Position
           for Ultraviolet Fluorometry

First of all, the photometer shutter must be closed to protect the PMT. The illuminator shutter in front of the UV lamp should be closed, but the secondary illuminator shutter in front of a halogen lamp should be open. The photometric field aperture should be out of the field of view, and a prompt may be needed for the operator to open the ocular shutter if previously closed.

3.4.5   Standardization of the Photometer
           for Fluorometry

The primary illuminator shutter in front of the UV lamp should be closed, but the secondary illuminator shutter in front of halogen lamp must be open. The photometric field aperture should be swung into the optical axis and properly centered. The operator may need a prompt to close the ocular shutter if necessary. Finally, the photometer shutter is opened.

3.4.6   Dark-Field

The photometer shutter should be closed, and there is no need to change any other shutters.

3.4.7   Dark-Field plus Ambient

Both primary and secondary illuminator shutters should be closed, and the photometric field aperture should be swung into line. The operator may need a prompt to open the ocular shutter, depending on the source of the ambient light under investigation. Finally, the photometer shutter is opened.

3.4.8   Negative Shutter Test

With the illuminator shutter open, the secondary illuminator shutter closed, the photometric field aperture in-line, the ocular shutter open or closed as appropriate, and the photometer shutter open, start a shutter test by closing and then opening the designated test shutter (using an initial graphics-frame delay and a specified length of time for the test shutter to remain closed). Shutter tests can be conducted using analog electronics as well as the digital method suggested above, and many photometer systems have an output for an oscilloscope or oscillograph.

3.4.9   Positive Shutter Test

With the illuminator shutter open, the secondary illuminator shutter closed, the photometric field aperture in-line, the ocular shutter open or closed as appropriate, and the photometer shutter open, start a shutter test by opening and then closing the designated test shutter (using an initial graphics-frame delay and a specified length of time for the test shutter to remain open).

3.4.10   Diagnostics

Open or close any shutter requested from the test menu, reporting on shutter status and GPIB service requests. Diagnostics can be called automatically as an error routine, using both ON TIMEOUT and ON ERROR statements at the top of SUB Shutter. Thus, all common errors related to shutters and apertures can be trapped and often solved without terminating the experiment. Trapping shutter errors can be very difficult if open and close commands for shutters are scattered through the main program. In general, swing-in solenoid shutters are extremely reliable, but the same cannot be said of all makes of high-speed camera shutters, especially if they have been exposed to an arc lamp.

3.4.11   Software

Programming the shutter control in a complex CAM may become rather confusing. If the operator has declared which shutter is being used for what function, it is possible to control the shutters globally. The commands passed into the subprogram listed below are (1) put shutters to the measuring position; (2) put shutters to a dark-field measuring condition; (3) put shutters to a viewing position; (4) protect the specimen from strong illumination; (5) expose the sample to strong illumination just before a measurement; (6) get ready to flap a shutter prior to making parallel measurements of a photometer response; (7) flap the shutter for the specified length of time; and (8) ensure the photometer shutter is open. The order in which the operations are undertaken is a reverse priority, so that the most important shutter movements are made last, thus superseding any earlier shutter state.

```
10     Shutter:SUB Shutter(Mpc(*),Command)
20     !
30     ASSIGN @Mpc1 TO 706 ! HP-IB address of one
       interface
40     ASSIGN @Mpc2 TO 708 ! HP-IB address of another
       interface
50     !
60     IF Mpc(20)=1 THEN DISP "in SUB Shutter, command
       =";Command
70     IF Mpc(18)=10 THEN ! fake shutter for fake PMT
80     Mpc(6)=-1 ! signal to SUB Pmt that fake shutter
       is shut
90     GOTO End_shutter
100    END IF
110    ON TIMEOUT 7,.5 GOTO 200 ! avoids fast-shutter
       lock-up
120    IF (Mpc(18)>0 AND Mpc(18)<5) OR (Mpc(22)>=1 AND
       Mpc(22)<=3) THEN ! Mpc interface #1 is in use
130    CLEAR @Mpc1 ! clear any previous errors
140    Mpcflag=1 ! internal flag for using first
       interface
150    END IF
160    IF Mpc(18)>5 OR (Mpc(22)>=4 AND Mpc(22)<=6) THEN
       ! the other MPC interface is in use
170    CLEAR @Mpc2 ! clear the second interface
180    Mpcflag=Mpcflag+2 ! internal flag
190    END IF
200    OFF TIMEOUT ! CIH, come in here, never erase
       this line
210    SELECT Command ! find what shutter configuration
       is needed
```

Mpc(18) holds the identity of the dark-field shutter, which will be opened, while the status of Mpc(19) will cause the photometric field aperture to be shut if it is in operation. For some values of Mpc(18) the control is to the first interface, @Mpc1, while others are routed through another interface, @Mpc2. There is also provision for a manual shutter and for a spare solenoid shutter (ILEX), usually operated when there is a miniature optical bench installed somewhere in the CAM.

```
220    CASE 1 ! shutters should be in position for a
       measurement
230    IF Mpc(18)=1 THEN OUTPUT @Mpc1 USING "#,A";"w"
       ! open swing-in
240    IF Mpc(18)=2 THEN OUTPUT @Mpc1 USING "#,A";"o"
       ! open fast shutter
```

```
250    IF Mpc(18)=4 THEN OUTPUT @Mpc1 USING "#,A";"v"
       ! open ILEX solenoid shutter
260    IF Mpc(19)=1 THEN OUTPUT @Mpc1 USING "#,A";"u"
       ! shut field stop
270    IF Mpcflag=1 OR Mpcflag=3 THEN OUTPUT @Mpc1
       USING "#,A";"q" ! open shutter under PMT
280    IF Mpc(18)=5 AND Mpc(21)<>1 THEN ! manual or
       fake shutter
290    BEEP 100,1 ! need to get operator's attention
       for this
300    PRINT
310    PRINT "open manual shutter, then CONT"
320    PAUSE
330    Mpc(21)=1 ! flag that manual shutter is open
340    END IF
350    IF Mpc(18)=6 THEN OUTPUT @Mpc2 USING "#,A";"w"
       ! open swing-in shutter
360    IF Mpc(18)=7 THEN OUTPUT @Mpc2 USING "#,A";"o"
       ! open fast shutter
370    IF Mpcflag=2 OR Mpcflag=3 THEN OUTPUT @Mpc2
       USING "#,A";"q" ! open shutter under PMT
380    IF Mpc(18)=9 THEN OUTPUT @Mpc2 USING "#,A";"v"
       ! open ILEX solenoid shutter
390    WAIT .1
```

A dark-field measurement requires a different set of operating conditions, primarily determined by the identity of the dark-field shutter given by Mpc(18):

```
400    CASE 2 ! shutters for a dark-field
410    IF Mpc(18)=1 THEN OUTPUT @Mpc1 USING "#,A";"v"
       ! shut swing-in shutter
420    IF Mpc(18)=2 THEN OUTPUT @Mpc1 USING "#,A";"n"
       ! shut fast shutter
430    IF Mpc(18)=3 THEN OUTPUT @Mpc1 USING "#,A";"p"
       ! shut MPM03
440    IF Mpc(18)=4 THEN OUTPUT @Mpc1 USING "#,A";"w"
       ! shut ILEX
450    IF Mpc(18)=5 AND Mpc(21)<>2 THEN ! manual or
       fake shutter in use
460    PRINT
470    BEEP 100,1 ! attract attention
480    PRINT "close manual shutter, then CONT"
490    PAUSE
500    Mpc(21)=2 ! flag manual shutter closed
510    END IF
520    IF Mpc(18)=6 THEN OUTPUT @Mpc2 USING "#,A";"v"
       ! shut swing-in shutter
530    IF Mpc(18)=7 THEN OUTPUT @Mpc2 USING "#,A";"n"
       ! shut fast shutter
```

```
540    IF Mpc(18)=8 THEN OUTPUT @Mpc2 USING "#,A";"p"
       ! shut MPM03
550    IF Mpc(18)=9 THEN OUTPUT @Mpc2 USING "#,A";"w"
       ! shut ILEX
560    WAIT .1
```

The main conditions for viewing are to illuminate the specimen and to remove the photometric field aperture which, otherwise, may hinder the operator from centering the specimen. The photometer shutter must be closed or else the PMT will be saturated.

```
570    CASE 3 ! need to view sample under microscope
580    ON TIMEOUT 7,1 GOTO Forgetit ! error trap for
       hardware
590    IF (Mpcflag=1 OR Mpcflag=3) AND Mpc(22)<>2 THEN
       OUTPUT @Mpc1 USING "#,A";"o"! open fast shutter
600    IF (Mpcflag=2 OR Mpcflag=3) AND Mpc(22)<>5 THEN
       OUTPUT @Mpc2 USING "#,A";"o"! open fast shutter
610    IF (Mpcflag=1 OR Mpcflag=3) AND Mpc(22)<>1 AND
       Mpc(18)<>4 THEN OUTPUT @Mpc1 USING "#,A";"w"
       ! open swing-in shutter
620    IF (Mpcflag=2 OR Mpcflag=3) AND Mpc(22)<>4 AND
       Mpc(18)<>9 THEN OUTPUT @Mpc2 USING "#,A";"w"
       ! open swing-in shutter
630    IF Mpcflag=1 OR Mpcflag=3 THEN OUTPUT @Mpc1
       USING "#,A";"t" ! open field stop
640    IF Mpcflag=2 OR Mpcflag=3 THEN OUTPUT @Mpc2
       USING "#,A";"t" ! open field stop
650    IF Mpcflag=1 OR Mpcflag=3 THEN OUTPUT @Mpc1
       USING "#,A";"p" ! shut shutter under PMT,
       protect PMT
660    IF Mpcflag=2 OR Mpcflag=3 THEN OUTPUT @Mpc2
       USING "#,A";"p" ! shut shutter under PMT,
       protect PMT
670    WAIT .1 ! some of these are shutters are slow
```

After typical operations such as looking for specimens to measure, and measuring blanks and specimens, it may be necessary to protect the specimen from fluorescence quenching or heat by closing the protective shutter for the specimen.

```
680    CASE 4 ! shut off strong light
690    IF Mpc(22)=1 THEN OUTPUT @Mpc1 USING "#,A";"v"
       ! shut swing-in shutter
700    IF Mpc(22)=2 THEN OUTPUT @Mpc1 USING "#,A";"n"
       ! shut fast shutter
710    IF Mpc(22)=3 THEN OUTPUT @Mpc1 USING "#,A";"w"
       ! shut ILEX
```

```
720    IF Mpc(22)=4 THEN OUTPUT @Mpc2 USING "#,A";"v"
       ! shut swing-in shutter
730    IF Mpc(22)=5 THEN OUTPUT @Mpc2 USING "#,A";"n"
       ! shut fast shutter
740    IF Mpc(22)=6 THEN OUTPUT @Mpc2 USING "#,A";"w"
       ! shut ILEX
750    PRINT "shutting off UV to sample"
760    WAIT .1 ! takes a while
```

A higher priority, however, is to open the protective shutter to make a measurement.

```
770    CASE 5 ! e.g., expose sample to UV to excite
       fluorescence
780    IF Mpc(22)=1 THEN OUTPUT @Mpc1 USING "#,A";"w"
       ! open slow shutter
790    IF Mpc(22)=2 THEN OUTPUT @Mpc1 USING "#,A";"o"
       ! open fast shutter
800    IF Mpc(22)=3 THEN OUTPUT @Mpc1 USING "#,A";"v"
       ! open ILEX solenoid shutter
810    IF Mpc(22)=4 THEN OUTPUT @Mpc2 USING "#,A";"w"
       ! open slow shutter
820    IF Mpc(22)=5 THEN OUTPUT @Mpc2 USING "#,A";"o"
       ! open fast shutter
830    IF Mpc(22)=6 THEN OUTPUT @Mpc2 USING "#,A";"v"
       ! open ILEX solenoid shutter
840    PRINT "exposing sample to UV"
```

An even higher priority will be if high-speed parallel measurements of the shutter operation are being studied, in which case the appropriate shutter must be shut before it is flapped open, or opened before it is flapped shut.

```
850    CASE 6 ! get ready to flap shutter
860    SELECT Mpc(24) ! close shutter ready for
       operation
870    CASE 1 ! want to examine operation of slow
       shutter
880    OUTPUT @Mpc1 USING "#,A";"v" ! shut swing-in
       shutter
890    CASE 2 ! want fast, segmented shutter
900    OUTPUT @Mpc1 USING "#,A";"n" ! shut fast
       shutter
910    END SELECT
920    WAIT .1 ! may have to wait for slow one
```

With the shutter poised for examination, it may now be opened or shut for the specified length of time, leaving the task of creating the x-axis time steps to the parallel measuring system.

```
930   CASE 7 ! flap shutter open for parallel
      measurements
940   IF Dark_flag=0 AND Mpc(24)=1 THEN OUTPUT @Mpc1
      USING "#,A";"w" ! open swing-in shutter
950   IF Dark_flag=0 AND Mpc(24)=2 AND Mpc(20)=1 THEN
      PRINT "opening fast shutter"
960   IF Dark_flag=0 AND Mpc(24)=2 THEN OUTPUT @Mpc1
      USING "#,A";"o" ! open fast shutter
970   WAIT Mpc(25) ! specified delay between opening
      & closing
980   IF Mpc(24)=1 AND Mpc(20)=1 THEN PRINT "shutting
      swing-in shutter" ! test message
990   IF Mpc(24)=1 THEN OUTPUT @Mpc1 USING "#,A";"v"
      ! shut swing-in shutter
1000  IF Mpc(24)=2 AND Mpc(20)=1 THEN PRINT "shutting
      fast shutter" ! test message
1010  IF Mpc(24)=2 THEN OUTPUT @Mpc1 USING "#,A";"n"
      ! shut fast shutter
```

The highest priority is to make sure that the photometer shutter has been opened before a measurement is made.

```
1020  CASE 8 ! ensure MPM03 is open so light can
      reach PMT
1030  IF Mpcflag=1 OR Mpcflag=3 THEN OUTPUT @Mpc1
      USING "#,A";"q" ! open shutter under PMT
1040  IF Mpcflag=2 OR Mpcflag=3 THEN OUTPUT @Mpc2
      USING "#,A";"q" ! open shutter under PMT
1050  END SELECT
1060  !
1070  End_shutter:IF Mpc(20)=1 THEN DISP "" ! clear
      test messages
1080  SUBEXIT
1090  !
1100  Forgetit:BEEP ! MPC was not there
1110  PRINT "an MPC box was inactive for a command,
      hit CONT"
1120  PAUSE
1130  SUBEND !-----------------------------------
```

3.5   Importance of Timing

To understand the importance of timing operations, consider the way in which the CAM has evolved. Some of the first affordable CAMs (such as the Zeiss Zonax) combined the controller with the interface in a single unit. But controller components soon became obsolete with PC advances, so that the investment in a controller-specific interface was wasted. Also, by combining the controller and the interface, it was difficult to take advantage of the competitive low pricing of mass-produced PCs. Thus, it became advantageous to separate the PC and the interface, linking them with a widely used bus such as the GPIB; however, when the user now became responsible for updating the PC, certain problems occurred. When a controller was updated, the new one always had a faster clock speed than the old one, so that software without explicit timing instructions (using implicit timing from the old controller) could fail on the new controller when instructions arrived at the interface more rapidly than they could be processed.

This software bug was very obvious if a device such as a monochromator stopped scanning altogether, but it easily escaped notice if the desynchronization was at a threshold level for the interface so that, occasionally and unpredictably, the monochromator missed a single step increment. The solution is to determine the length of time required for each operation and to write this into the software — this is the great advantage of incorporating a multiprogrammer into the system. The logic for a scanning monochromator is the same as that illustrated for the simple case of a shutter but is tedious because of the programming required for all the different wavelengths. Once having acquired and installed a multiprogrammer, it makes sense to use the multiprogrammer to enhance the performance of the CAM for an endless variety of new applications.

Chapter 4

Photometry

4.1 Introduction

Most of the operations of the CAM revolve around photometry. Photometry is a monumentally large topic, ranging all the way from Bunsen's grease-spot photometer to the Hubble space telescope. We will be working somewhat closer to the grease-spot than to the Hubble, starting with simple photometers that can turn an LM into a CAM on a really low budget, then working up to the PMT, which still is the dominant technology for the CAM. Thus, older references, such as Gibson (1949), may be of more use than the current literature on charge-coupled device (CCD) photometers.

4.2 Analog-to-Digital Conversion

4.2.1 Dual-Slope Method

The multiplicity of ADC methods makes it difficult to generalize. Dual-slope ADCs use the current of the test voltage through a known resistor to drive an integrating amplifier that charges up a capacitor which, after a set time, is completely discharged by a reference current. Thus, integration of the two slopes yields a ratio of the test current to the reference current. Being derived from a ratio, the results are reliable, and 60-Hz-line power-supply interference may be canceled by ensuring the integration times involve complete AC cycles (so that the positive part of the cycle cancels the negative part of the cycle during integration); however, the dual-slope method is relatively slow.

4.2.2   Successive Approximation Method

Faster methods use a voltage-to-frequency converter, integrating the input voltage to a series of pulses, which then can be secondarily integrated by counting over a period of time. But the favorite PC method is the successive approximate ADC. The register of an accurate digital-to-analog converter (DAC) is tested, one bit at a time, from the most significant bit (MSB) to the least significant bit (LSB). If the MSB produces an output less than the test voltage, it is kept high (1), but if the output is lower than the test voltage, it is made low (0). Then the next MSB is tested, and so on, down to the LSB. Thus, the conversion time is proportional to the number of bits tested. This sequence of operations also can be done in parallel by massive replication of the circuits, as in the flash ADCs used for real-time ADC of voice and images.

4.2.3   Nyquist Theorem

ADC speed is seldom critical for the type of CAM operations described in this book. But it is important at least to be aware of one of the major practical implications of the Nyquist theorem, that the rate of data acquisition must exceed twice the highest frequency present in the signal. With normal operations of the CAM, most of which involve integrating damped signals or using numerous data points to outline relatively slow peaks in the signal, there is seldom a problem. But, with strong interference superimposed on a raw signal there could be. If the rate of sampling is a harmonic of the interference, and if the integration period happens to correspond to an ascending (constructive) or descending (destructive) phase of the aliased signal, then a surprisingly large positive or negative bias is produced. Thus, if the sampling rate happens to correspond to the peaks of any regular interference, the interference peaks are added to the signal (because the valleys, which would normally cancel the peaks, have been missed).

There are two ways to approach CAM ADCs. With lots of technical talent and a low budget, the DAC registers can be programmed directly from an output port of the PC, feeding the logic output of a comparator amplifier (comparing the DAC output to the test voltage) into a PC input port. With medium technical talent and medium budget, one's time is best spent programming an ADC card that is accessible on the GPIB, usually a fully programmable 12-bit ADC, evaluating performance by the quality of the sample-and-hold amplifier (the shortness of sampling time together with the reliability of holding the result to the end of the cycle). There might be a third option, no technical talent and a big budget, but one would not expect administrators to be reading this book.

4.3   Photoresistive Detectors

It is doubtful whether any commercial CAMs use photoresistive detectors for photometry, but this is where a student on a minimal budget might start, replacing one

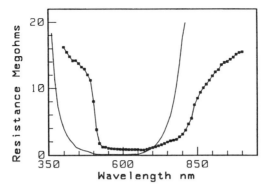

FIGURE 4.1
Relative spectral responses of two different photoresistors in a CAM.

ocular in a binocular with a photoresistor and using the field aperture as the photo-metric aperture. Using a low-power objective and a powerful illuminator, one can make reliable, simple measurements and develop an appreciation of the engineering that goes into a commercial CAM.

Light energy may remove electrons from their parent atoms in a photoresistor, so that the free electrons carry a current which reduces overall resistance. As in using the LM to make photomicrographs with a camera, the best way to manipulate light intensity is by using neutral density (gray) filters, because reducing the illuminator voltage changes the emission spectrum of the illuminator (color balance is lost in photography, and spectrophotometry data are biased). The resistance of a CAM photoresistor increases when a neutral density filter is inserted.

4.3.1   Spectral Response

On measuring equal areas of more or less uniformly stained structures, it is soon discovered that the spectral sensitivity of the photoresistor is unlikely to match that of the human eye. Thus, red structures appearing dark to the eye may register a high transmittance with the photoresistor. Figure 4.1 shows a comparison of two hobby-shop photoresistors, both responsive to infrared light not seen by eye — which explains why both photoresistors responded strongly to red light at 700 nm which is only just visible. Thus, when using a photodetector to make a CAM measurement, the most important step is the most simple one: making sure that the photodetector responds to the wavelengths of interest. The lowest levels of response at low and high wavelengths should be avoided, if possible, so that the working range should be across regions of the spectrum to which the detector responds adequately. Also, from Chapter 1, remember that LM objectives are only corrected (with varying degrees of effectiveness) for chromatic aberration of visible light. With Köhler illumination established, and everything in the light path (slide, mounting medium, cover slip, and immersion oil, if used) except the specimen, the operator may scan across the working range of the spectrum to find the photometer response at each wavelength. These values are used as the 100% transmittance values, to which actual measurements of the specimen are compared. Before the days of computer assistance, these values had to be recorded manually, so that even calculating a simple transmittance spectrum was rather tedious. Even more tedious than the transmittance spectrum was the absorbance spectrum. Absorbance is the common logarithm of the reciprocal of transmittance. With computer operation, the 100% transmittance values are stored on disk (as well as in a COM block, if possible). Rather than storing a simple vector, it is useful to ALLOCATE a matrix that can handle a variety of similar data related to each wavelength (dark-field current, fluorescence blank, etc.). A flexible approach to programming is needed because the number of wavelengths may differ from one experiment to the next.

4.4   Photovoltaic Detectors

Silicon atoms in a semiconductor share their four valence electrons in covalent bonds with other atoms. The presence of gallium or indium atoms with three valence electrons creates the p-zone of the photodetector, so called because a positive excess charge is caused by the missing negatively charged electrons (electron holes). Phosphorus, antimony, or arsenic with five valence electrons, on the other hand, may create the n-zone, which is negatively charged because of the extra electrons. The PIN-diode has an i-zone sandwiched between the p- and n-zone layers. The i-zone, also called the intrinsic-zone or depletion region, has a low electrical conductivity until a photon with the appropriate energy is absorbed. The energy of the photon is related to its wavelength, and the energy gap required depends on the chemical composition of the i-zone. Absorption of the photon creates a hole-electron pair. The hole moves to the anode in contact with the p-zone, while the electron moves to the cathode in contact with the n-zone, producing a detectable current flow. In a typical communications PIN-diode, the photon arrives through the p-zone, which is kept thin, and the i-zone is relatively thick to increase sensitivity. Rapid response is gained from a reverse electrical bias imposed on the PIN-diode.

4.4.1   Solar Cell

In the CAM, if speed of response is relatively unimportant, a solar cell can be used as a photodetector. The i-zone of the PIN-diode configuration is now thin, low-resistance materials are used, and a current proportional to light intensity will flow through an external circuit. Figure 4.2 shows the spectral sensitivity of a solar cell mounted in a CAM for use at low magnification with a strong illuminator and large photometric aperture. The solar cell is relatively slow, because of the capacitance of the thin i-zone, but mechanical scanning with a monochromator is generally a slow operation anyway.

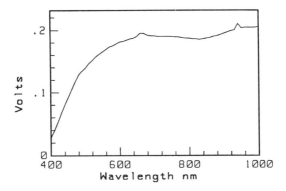

FIGURE 4.2
Relative spectral response of a solar cell in a CAM.

4.5   Photomultipliers

Using a PMT under computer control is well documented (Zubchenok et al, 1988; Devlin et al., 1988; Bretagnon et al., 1992; Williams et al., 1992; Stamm et al., 1993) and well supported by commercial apparatus. In a CAM, the PMT is used for a wide range of applications, such as spectrophotometry, spectrofluorometry, and ellipsometry, and it can cope with a great range in light intensities if used properly. If a PMT is refrigerated and never exposed to strong light, as in scintillation counting, it has advantageous features such as rapid response and linearity (Johnson, 1970; Knoll, 1979). But in the less protected environment provided by a CAM, the PMT signal may become noisy and require protection. Protection can be hard-wired (Ermakov et al., 1986) or achieved by cautious programming of the dynamic range of the PMT (Swatland, 1996a), as described here.

4.5.1   Hardware

Although two main types of PMT are available, with a window at either the side or end of the vacuum tube, the side-window PMT is used most often on the CAM because it is smaller and less expensive. The transmittance of the window in the vacuum tube imposes a limitation on the minimum wavelength of photons able to impinge on the photocathode, and very few CAMs can use light much below 320 nm. Photons striking the photocathode immediately inside the window cause the photoemission of electrons which are focused by an electrostatic field and accelerated towards the first dynode. Each high-energy electron causes the emission of two or more secondary electrons which, in turn, accelerate towards the next dynode, which is at a higher positive voltage than the last. An anode, in front of the last dynode with the highest voltage in the series, collects the current. Side-window PMTs in a CAM are normally used to a maximum of approximately 1.2 kV. The dark-field current from thermionic emission at the cathode may become quite substantial at high voltage.

4.5.2   Software

Simple power-up loop programming instructions for PMT controllers are readily available, as in Zeiss Bulletin G41-912-e:

```
10     SUB Set_pmt! whose address should be assigned
       to @Path
20     !
30     ASSIGN @Path TO 706
40     PRINT "create maximum light condition"
50     PRINT "then enter the gain (10,100 or 1000)",
60     INPUT Gain
70     PRINT Gain
80     SELECT Gain
90     CASE 10
100    OUTPUT @Path USING "#,A";"B" ! × 10
110    CASE 100
120    OUTPUT @Path USING "#,A";"D" ! × 100
130    CASE 1000
140    OUTPUT @Path USING "#,A";"H" ! × 1000
150    END SELECT
160    OUTPUT @Path USING "#,A";VAL$(0) ! no
       integration
170    OUTPUT @Path USING "#,A";"R" ! select lowest
       voltage
180    WAIT 2 ! takes a while
190    INTEGER A
200    Reps=10 ! number of replicates
210    Start: Pmt=0 ! clear register
220    FOR I=1 TO Reps ! for each replicate
230    OUTPUT @Path USING "#,A";"!" ! start ADC
240    ENTER @Path USING "#,W";M ! enter result
250    Pmt=Pmt+M ! add result to stash
260    NEXT I
270    Pmt=Pmt/Reps ! average result
280    PRINT "PMT = ";Pmt, ! show it
290    A=SPOLL(@Path) ! look for PMT saturation
300    PRINT "poll = ";A, ! show state
310    IF BIT(A,6) AND BIT(A,4) THEN BEEP ! sound
       alarm
320    !
330    IF M<3000 THEN ! increase voltage
340    OUTPUT @Path USING "#,A";"h" ! increase
350    Hv=Hv+1 ! keep count of increase
360    PRINT " high voltage = ";Hv
```

```
370     GOTO Start
380     END IF
390     !
400     IF M>3700 AND Hv>1 THEN ! decrease voltage
410     OUTPUT @Path USING "#,A";"l" ! decrease
420     Hv=Hv-1 ! update register
430     PRINT "high voltage = ";Hv*3.5 ! convert
        counter to volts
440     GOTO Start
450     END IF
460     !
470     Max=0 ! clear register
480     FOR I=1 TO Reps ! for each replicate
490     OUTPUT @Path USING "#,A";"!" ! start ADC
500     ENTER @Path USING "#,W";M ! enter result
510     Max=Max+M ! add value to stash
520     NEXT I
530     Max=Max/Reps ! find average
540     PRINT "final PMT output = ";Max
550     !
560     SUBEND!------------------------
```

A simple subprogram such as this could be installed anywhere it was needed. But, as in the case of the shutters and apertures considered in the previous chapter, in the long run there are many advantages to controlling all PMT operations in a subprogram that can be called from any main program to perform any photometric function. Thus, radical changes made to the hardware of the CAM can be handled without further programming, handling substitutions interactively rather than by reprogramming. This can be accomplished by holding a set of operating flags and registers in an array declared as a COM block in the main program, as shown below. The COM block survives termination of the program, so that the PMT can be kept alive between program runs. Without this, or some equivalent file surviving, the PMT would have to be reset each time the program is run for the program to contain data on dark-field outputs, maximum values, etc.

```
10      Pmt:SUB Pmt(Mpc(*))
20      !
40      IF Mpc(6)<>0 AND Mpc(6)<>1 AND Mpc(6)<>2 AND
        Mpc(6)<>3 AND Mpc(6)<>4 AND Mpc(6)<>5 AND
        Mpc(6)<>10 AND Mpc(6)<>11 THEN ! need to know
        if called incorrectly from new prgm
50      BEEP
60      ALPHA ON
70      PRINT "CAME INTO SUB Pmt WITHOUT VALID
        INSTRUCTION"
80      PAUSE
90      SUBEXIT
100     END IF
```

```
110   CALL Hp(Id) ! see what sort of controller is
      running prgm
120   IF Mpc(20)=1 THEN DISP "in SUB Pmt" ! only if
      testing
```

In the few lines that follow, there are codes to handle different types of photometers, ranging from a simple photoresistor measured via a bridge circuit to a high-speed ADC card following the PMT output. The null photometer is a keyboard entry or a random number used in creating a fake data set to test statistical subprograms. If the photometer is a photodiode array (PDA, as considered in Chapter 10), the output of the subprogram is a vector rather than a scalar and cannot easily be pre-defined. Thus, when invoked by the code for a PDA, the result of the photometer measurement is to be found by the main program, not within the array Mpc(*), which requires fixed dimensions defined at the start of all main programs, but in another array of variable dimensions. This allows the photometer to produce a vector rather than a scalar and is useful in many ways. For example, the photometer output might be a curve describing the initial response of a PMT to a pulse of light, as described in Chapter 3.

```
130   SELECT Mpc(14) ! look for non standard
      photometer
140   CASE 1 ! want International Light IL700A
      Radiometer
150   CALL Radio(Mpc(*)) ! get measurement from IL700A
160   GOTO Pmt_end ! then exit
170   CASE 2 ! using a photoresistor
180   CALL R_photo(Mpc(*)) ! read from resistor bridge
190   GOTO Pmt_end ! then exit
200   CASE 3 ! using high-speed ADC card to follow
      PMT response
210   IF Mpc(0)>0 THEN ! only if system already
      initialized
220   CALL Flap_pmt(Mpc(*)) ! get vector
230   GOTO Pmt_end ! then quit
240   END IF
250   CASE 4 ! fake PMT for debugging new system or
      stats testing
260   IF Mpc(6)=0 THEN PRINT "initialize fake PMT" !
      pretend
270   IF Mpc(6)=4 THEN PRINT "power-down fake PMT" !
      do it
280   IF Mpc(6)=11 THEN Mpc(11)=RND ! random number
290   IF Mpc(6)=3 OR Mpc(6)=0 THEN ! fake initialize
      or read
300   PRINT "fake PMT reading" ! remind user of fake
      status
310   Mpc(8)=1 ! fake PMT reading at standardization
320   Mpc(9)=.01 ! fake dark-field reading at
      standardization
330   Mpc(11)=1 ! fake most recent PMT reading
```

```
340    END IF
350    IF Mpc(6)=2 THEN Mpc(9)=.01 ! fake read dark
       field only
360    IF Mpc(6)=1 THEN Mpc(11)=.02 ! fake PMT reading
370    PRINT "fake data"; Mpc(8), Mpc(9), Mpc(11)
380    GOTO Pmt_end ! exit after faking data
390    END SELECT ! end of non-Zeiss photometers
400    IF Mpc(14)=0 OR Mpc(14)=3 THEN ! use interface
       MPC#1
410    ASSIGN @Mpc TO 706 ! HP-IB address
420    IF Mpc(20)=1 THEN PRINT "using MPC#1 at 706" !
       test
430    END IF
440    IF Mpc(14)=5 THEN ! use Zeiss PMT via interface
       MPC#2
450    ASSIGN @Mpc TO 708 ! HP-IB address
460    IF Mpc(20)=1 THEN PRINT "using MPC#2 at 708" !
       test
470    END IF
480    GOSUB Default ! check all required operating
       conditions
490    ALLOCATE One(0,(Mpc(3)*10))! used for
       stabilization routine
```

Now follows the start of the SUB's main control program, where the instruction passed into the subprogram as Mpc(6) determines what operation is undertaken. This cannot be done logically unless the dark-field shutter has been properly identified. As described in Chapter 3, this is a critical decision for the operator because it can have a major effect on performance. In many situations, a constant low level of ambient illumination may be regarded as part of the dark-field output of the photometer (to be subtracted from each measurement), but it is always wise to check that the ambient illumination is both low and constant. Typical sources of error arise from ambient illumination reflected into the optical axis of the microscope from the operator's arm in a white laboratory coat, overhead room light reaching the photometer when the operator's head is moved away from the viewing position, fluorescent room lights flickering at the frequency of their power supply, stray light from a nearby computer screen and panel-mounted diodes, and cloud movement across a window.

From the viewpoint of software universality, it is essential to have a standard protocol for subtracting the dark-field current from a measurement. It matters little where it is done, as long as it is always done, only once, and only in the right place. Perhaps it is safest never to make this subtraction in a subprogram, but always to do it in the main program where it is conspicuous.

Running SUB Pmt updates a variety of memory registers, which are available in the main program. The amplifier memory is the degree of amplification for the PMT after it was last changed, whereas the amplifier target is the required gain when the instruction is to reset an already operational PMT. This allows the photometer to be re-set under program control, loading the target from an array in the main program, as in the duplication of previously existing conditions. The PMT voltage memory and

target are used for the same purpose. Amplifier gain and voltage at time of last
standardization may be used to assess drifting by re-measuring a standard under
program control after a series of critical measurements.

```
510   IF Mpc(0)=0 OR Mpc(6)=0 THEN GOSUB Pmt_first !
      initialize
520   IF Mpc(6)=0 OR Mpc(6)=5 THEN ! normal
      measurement
530   CALL Shutter(Mpc(*),1) ! shutters to measuring
      position
540   GOSUB Pmt_stand ! restart
550   END IF
560   IF Mpc(6)>=10 THEN GOSUB Reset! reset targets
570   IF Mpc(6)=1 OR Mpc(6)=3 THEN GOSUB Measure_pmt
      ! normal
580   IF Mpc(6)=2 OR Mpc(6)=3 THEN GOSUB Dark !
      dark-field
590   IF Mpc(6)=4 THEN ! instruction is power down
600   OUTPUT @Mpc USING "#,A";"R" ! set PMT to 0
      volts
610   Mpc(2)=0 ! reset register for high voltage
      memory
620   ALPHA ON ! show alphanumerics on screen
630   BEEP ! attract attention
640   PRINT "SUB Pmt powering-down PMT, exogenous
      command"
650   WAIT 1 ! allow voltage to drop
660   END IF
670   GOTO Pmt_end ! exit normally
680   !
690   !.................SUBROUTINES..................
700   !
710   Default: ! set default values
720   !
730   IF Mpc(3)<=0 THEN ! forgot to state how many
      PMT replicates
740   Mpc(3)=20 ! default value
750   PRINT "number of PMT reps defaulted to";Mpc(3)
760       END IF
770   IF Mpc(4)<>10 AND Mpc(4)<>100 AND Mpc(4)<>250
      AND Mpc(4)<>500 AND Mpc(4)<>750 AND Mpc(4)<>1000
      AND Mpc(4)<>5000 THEN ! forgot to state damping
      required
780   Mpc(4)=10 ! default is strong damping of signal
790   PRINT "damper defaulted to";Mpc(4);" Hz"
      ! show operator
800   END IF
810   IF Mpc(10)<=0 THEN ! forgot to state dynamic
      range
820   Mpc(10)=.8 ! enough to be useful
```

```
830   PRINT "fraction of dynamic range defaulted
      to";Mpc(10)
840   END IF
850   IF Mpc(16)<=0 THEN ! forgot degree of
      stabilization
860   Mpc(16)=2 ! 2% is default
870   PRINT "stabilization % defaulted to";Mpc(16)
880   END IF
890   IF Mpc(17)<=0 THEN ! forgot size of acceptance
      window
900   Mpc(17)=.05 ! medium size default
910   PRINT "half window defaulted to";Mpc(17)
920   END IF
930   IF Mpc(18)<=0 THEN
940   PRINT ! CIH, come in here, so never remove this
      line
950   PRINT "identify the dark-field shutter" !
      operator input
960   PRINT "-------------------------------"
970   PRINT
980   PRINT "IF YOU ASKED FOR A FLUORESCENCE OPTION,
      THEN THE"
990   PRINT "DARK-FIELD SHUTTER IS IN THE PATH OF THE
      HALOGEN"
1000  PRINT "SECONDARY SOURCE FOR SETTING PMT TO
      WHITE LIGHT."
1010  PRINT
1020  PRINT "-----------------------------------------"
1030  PRINT "for slow 46-72-26 swing in shutter......
      enter 1"
1040  PRINT "for fast 46-72-25 fast segment
      shutter........ 2"
1050  PRINT "for shutter in MPM03 photometer
      head.......... 3"
1060  PRINT "for ILEX shutter in miniature optical
      bench... 4"
1070  PRINT "for manual shutter.................. 5",
1080  INPUT Mpc(18) ! input dark-field shutter
      identity
1090  IF Mpc(18)<>1 AND Mpc(18)<>2 AND Mpc(18)<>3 AND
      Mpc(18)<>4 AND Mpc(18)<>5 THEN 940
1100  PRINT Mpc(18) ! show it
1110  IF Mpc(18)=5 THEN Mpc(47)=1 ! manual shutter
      flag
1120  END IF
1130  IF Mpc(23)=0 THEN PRINT "low precision
      measurements"
1140  IF Mpc(23)=1 THEN PRINT "high precision
      measurements"
1150  RETURN ! finished setting default values
```

There are many little checks to be made when a CAM is first powered up and operating variables that must be declared or estimated. Each component of the CAM has its own initialization flag, Mpc(0) in this example. When set at 0, it called Pmt_first from line 510, which requested the operator to check appropriate hardware, then the flag is set to 1. The status of the flag is maintained in a COM block which survives until the controller is powered down or reset. The initialization routine also can be called by the main program as part of an error trap.

```
1190  Pmt_first: ! initialize: set PMT first time
1200  !
1210  IF Mpc(5)=1 THEN ! anticipate low light
1220  OUTPUT @Mpc USING "#,A";"H" ! set PMT to
      × 1000
1230  Mpc(1)=1000 ! amplifier gain
1240  END IF
1250  IF Mpc(5)=2 THEN ! anticipate medium light
1260  OUTPUT @Mpc USING "#,A";"D" ! set PMT to × 100
1270  Mpc(1)=100! memory of amplifier gain
1280  END IF
1290  IF Mpc(5)=0 OR Mpc(5)=3 THEN ! anticipate high
      light
1300  OUTPUT @Mpc USING "#,A";"B" ! set PMT to × 10
1310  Mpc(1)=10! memory of amplifier gain
1320  END IF
1330  IF Mpc(4)=10 THEN OUTPUT @Mpc USING "#,A";"6" !
      set 10 Hz
1340  IF Mpc(4)=100 THEN OUTPUT @Mpc USING "#,A";"5"
      ! or 100 Hz
1350  IF Mpc(4)=250 THEN OUTPUT @Mpc USING "#,A";"4"
      ! or 250 Hz
1360  IF Mpc(4)=500 THEN OUTPUT @Mpc USING "#,A";"3"
      ! or 500 Hz
1370  IF Mpc(4)=750 THEN OUTPUT @Mpc USING "#,A";"2"
      ! or 750 Hz
1380  IF Mpc(4)=1000 THEN OUTPUT @Mpc USING "#,A";"1"
      ! or 1 kHz
1390  IF Mpc(4)=5000 THEN OUTPUT @Mpc USING "#,A";"0"
      ! or 5 kH
1400  OUTPUT @Mpc USING "#,A";"R" ! high voltage off
1410  Mpc(2)=0 ! reset register for high voltage
      memory
1420  ALPHA ON ! see alphanumerics
1430  PRINT "waiting 5 sec for HV to drop"
1440  WAIT 5 ! this seems long enough
1450  ALPHA OFF ! clear screen to watch volts rise
1460  OUTPUT @Mpc USING "#,A";"h" ! set 320 V
1470  WAIT .5 ! this takes a little while
1480  Mpc(0)=1! only do this once
```

```
1490  IF Bufsiz=0 THEN GOSUB Framit ! graphics frame
1500  X(Tally)=320 ! lowest voltage, will appear on
      screen
1510  Y(Tally)=0 ! no PMT response at this voltage
1520  GOSUB Plotit ! update screen graphics as
      operator watches
1530  RETURN ! finished initialization
```

A number of important concepts appear in the following graphics subroutine. Essentially, a window or box is drawn, and the operator will watch the PMT output rise until it ends within the window. One of the determinants of the size of the box is the fraction of dynamic range of the PMT to be used. When initializing a PMT, a decision must be made on how much of the dynamic range to use. If the full dynamic range is used (fraction = 1), then any noise may cause the PMT to saturate. If only the lowest part of the dynamic range is used (say, 0.1), then the PMT is not being used to full effect. The default might be a compromise between protecting the PMT while not losing too much dynamic range, such as 0.8. Whereas the fraction of the dynamic range sets the height of the middle of the window, the vertical size of the window itself is determined by another variable (half-acceptance window).

The simplest approach to powering-up a PMT is to increase the voltage until a suitable PMT output is reached. However, this is not robust when light intensities are low because, as the noise increases relative to the signal, the safety margin between noise superimposed on the signal vs. the saturation point for the PMT becomes risky. Thus, instead of increasing the voltage until the PMT output ≥ (maximum output × dynamic range fraction), it is safer to specify an acceptance window above and below the dynamic range fraction so that any overshoot by the PMT can be detected, especially when the dynamic range fraction approaches 100%. Thus, the photometer output is accepted when it is > (maximum output × (dynamic range fraction − acceptance window)) and when the photometer output is < (maximum output × (dynamic range fraction + acceptance window)).

```
1550  Framit: ! set up screen graphics for operator
      to watch PMT
1560  Tally=6 ! 1 to 5 used for acceptance window
1570  IF Bufsiz>0 THEN ! this should not happen, but
      who knows?
1580  BEEP
1590  PRINT "CAME INTO SUB Framit WITH Bufsiz ALREADY
      SET"
1600  PAUSE
1610  RETURN
1620  END IF
1630  Bufsiz = 1000 ! enough resolution to get smooth
      lines
1640  ALLOCATE Leg$(5)[50],X(Bufsiz),Y(Bufsiz),
      Dotter(Bufsiz)
```

```
1650  Leg$(2)=VAL$(Mpc(4))&"Hz"&"  n="&VAL$(Mpc(3))&"
      DR"&VAL$(Mpc(10))&"  Esc"&VAL$(Mpc(16))&"%
      HW"&VAL$(Mpc(17))&"  HP"&VAL$(Mpc(23))  ! top
      title
1660  Leg$(3)="Volts" ! x-axis title
1670  Leg$(4)="PMT Output"! y-axis title
1680  GINIT ! initialize graphics
1690  SEPARATE ALPHA FROM GRAPHICS ! not for HP BASIC
      on IBM
1700  PLOTTER IS 3,"INTERNAL" ! use screen as plotter
1710  GRAPHICS ON ! not suitable for IBM windows
1720  ALPHA OFF ! this is running on an HP
      work-station
1730  X_gdu_max=100*MAX(1,RATIO) ! graphics display unit
1740  Y_gdu_max=100*MAX(1,1/RATIO) ! and another for
      the y-axis
1750  LORG 6 ! main title orientation is horizontal
1760  MOVE X_gdu_max/2,Y_gdu_max ! go to midpoint
1770  LABEL Leg$(2) ! use coded label
1780  DEG ! for y-axis title, it will be oriented in
      degrees
1790  LDIR 90 ! vertical label
1800  MOVE 0,Y_gdu_max/2 ! go to midpoint
1810  LABEL Leg$(4) ! put in y-axis label
1820  LORG 5 ! for x-axis title
1830  LDIR 0 ! horizontal
1840  MOVE X_gdu_max*.57,.15*Y_gdu_max ! goto midpoint
1850  LABEL Leg$(3) ! label x-axis
1860  VIEWPORT .15*X_gdu_max, .95*X_gdu_max,
      .25*Y_gdu_max, .9*Y_gdu_max
1870  FRAME ! want frame around viewport
1880  Xmin=300 ! lowest voltage is 320 V after reset
1890  Xmax=320+(255*3.5) ! reset 320 V + 255 steps of
      3.5 v
1900  Ymin=0 ! minimum PMT response
1910  Ymax=4100 ! should not get this high
1920  X(1)=Xmin ! mark acceptance window, the
1930  X(2)=Xmax ! vector showing the voltage rise
      also will be
1940  X(3)=Xmax ! used to draw a box which shows
      where the
1950  X(4)=Xmin ! acceptance window is located,
1960  X(5)=X(1) ! when waiting for PMT to end in box
1970  Y(1)=4000*(Mpc(10)-Mpc(17)) ! end box size
1980  Y(2)=Y(1) ! also for box
1990  Y(3)=4000*(Mpc(10)+Mpc(17)) ! also for box
2000  Y(4)=Y(3) ! also for box
2010  Y(5)=Y(1) ! also for box
2020  Dotter(1)=11 ! dotter is a vector for linetype
      control
```

```
2030 Dotter(2)=11 ! what line type to use
2040 Dotter(3)=11
2050 Dotter(4)=11
2060 Dotter(5)=11
2070 Xtick=200 ! size of graphics tick marks on
     x-axis
2080 Ytick=1000 ! size of graphics tick marks on
     y-axis
2090 WINDOW Xmin,Xmax,Ymin,Ymax ! window of what we
     can see
2100 AXES Xtick,Ytick,Xmin,Ymin,Xtick*10,Ytick*10,5
2110 CLIP OFF ! disable soft clip limits
2120 FOR I=Xmin TO Xmax STEP Xtick ! along x-axis
2130 LORG 6! places of number by tick mark
2140 MOVE I,Ymin-((Ymax-Ymin)/50)! under x-axis
2150 LABEL USING "#,K";I ! # suppress EOL
2160                     ! , is separator
2170                     ! K makes it compact
2180 NEXT I
2190 LORG 8 ! places number by tick mark
2200 FOR I=Ymin TO Ymax STEP Ytick ! along y-axis
2210 MOVE Xmin-((Xmax-Xmin)/60),I ! move to right
     place
2220 LABEL USING "#,K";I ! label as before
2230 NEXT I
2240 FOR Preplot=1 TO 4 ! draw acceptance window
2250 PLOT X(Preplot),Y(Preplot)
2260 NEXT Preplot
2270 PENUP
2280 CLIP ON ! everything now should be inside
     window
2290 RETURN ! ready to start plotting
2300 !
2310 Plotit: ! set line type to indicate amplifier
     gain
2320 IF Mpc(1)=10 THEN Dotter(Tally)=1 ! white line
     is × 10
2330 IF Mpc(1)=100 THEN Dotter(Tally)=2 ! red line
     is × 100
2340 IF Mpc(1)=1000 THEN Dotter(Tally)=3 ! yellow
     line is × 1000
2350 IF Id=360 OR Id=362 THEN ! these controllers
     plot color
2360 PEN Dotter(Tally)
2370 ELSE ! but the rugged monochrome controller
     does not
2380 LINE TYPE Dotter(Tally) ! so change the dotting
     pattern
2390 END IF
2400 PLOT X(Tally),Y(Tally) ! plot the PMT response
```

```
2410  Tally=Tally+1 ! increment the counting tally
2420  IF Tally=Bufsiz THEN Tally=6 ! loop back to
      start if
2430  RETURN ! have finished plotting
2440  !
2450  !.....................
2460  !
2470  Pmt_stand: ! standardize the PMT, can start
      with any values
2480  !
2490  A=0 ! A gathers data (B) from PMT
2500  FOR I=1 TO Mpc(3) ! Mpc(3) = number of
      replicates
2510  OUTPUT @Mpc USING "#,A";"!" ! cycle ADC
2520  ENTER @Mpc USING "#,W";B ! get data
2530  A=A+B ! add it in to stash
2540  NEXT I
2550  A=A/Mpc(3) ! get mean
2560  IF Mpc(6)=0 OR Mpc(6)=5 THEN ! only if
      initializing
2570  IF Bufsiz=0 THEN GOSUB Framit ! need new graphics
2580  X(Tally)=320+(Mpc(2)*3.5) ! voltage for graphics
      x-axis
2590  Y(Tally)=A ! PMT response for y-axis
2600  GOSUB Plotit ! plot the response and voltage
2610  END IF
2620  IF A>4090 THEN ! should never get here, but who
      knows?
2630  OUTPUT @Mpc USING "#,A";"R" ! power off as soon
      possible
2640  Mpc(2)=0 ! reset voltage memory
2650  ALPHA ON ! interrupt any graphics plotting
2660  PRINT "waiting 5 sec to power down PMT"
2670  WAIT 5
2680  ALPHA OFF ! back to graphics
2690  OUTPUT @Mpc USING "#,A";"h" ! set 320 V
2700  WAIT .5 ! takes a while
2710  IF Bufsiz=0 THEN GOSUB Framit ! may need
      graphics frame?
2720  X(Tally)=320 ! minimum value for x-axis
2730  Y(Tally)=0 ! minimum value for y-axis
2740  GOSUB Plotit ! plot the new values
2750  BEEP ! warn operator who may have missed this
      emergency
2760  ALPHA ON ! need to see on screen
2770  IF Sat_flag=5 THEN ! count number of
      saturations
2780  PRINT "REDUCE LIGHT INTENSITY AND START AGAIN"
2790  PRINT "-------------------------------------"
2800  BEEP 200,10 ! a different audible warning to normal
```

```
2810  PRINT "hit CONT when ready"
2820  PAUSE
2830  END IF
2840  PRINT "PMT SATURATED ----> RESET" ! update
      operator
2850  WAIT 5 ! takes a while
2860  IF Sat_flag>0 THEN ! has happened before
2870  IF Mpc(1)=100 THEN ! decrease amplifier
2880  Mpc(1)=10 ! was × 100, now is × 10
2890  OUTPUT @Mpc USING "#,A";"B" ! reset the gain
2900  END IF
2910  IF Mpc(1)=1000 THEN ! decrease amplifier
2920  Mpc(1)=100 ! was × 1000, now is × 100
2930  OUTPUT @Mpc USING "#,A";"D" ! reset the gain
2940  END IF
2950  END IF
2960  Sat_flag=Sat_flag+1 ! count the number of
      saturations
2970  ALPHA OFF ! clear screen for graphics
2980  GOTO Pmt_stand ! restart
2990  END IF ! end emergency response to saturated
      PMT
3000  !
3010  !.............
3020  !
3030  ! increase amplifier gain because early PMT
      response is low
3040  IF Mpc(2)=10 AND A<20 AND Mpc(1)=10 THEN !
      A < 20 is low
3050  OUTPUT @Mpc USING "#,A";"D" ! × 100
3060  Mpc(1)=100 ! memory of amplifier gain
3070  GOTO Pmt_stand ! restart
3080  END IF
3090  IF Mpc(2)=10 AND A<20 AND Mpc(1)=100 THEN
3100  OUTPUT @Mpc USING "#,A";"H" ! gain is now
      × 1000
3110  Mpc(1)=1000 ! memory of amplifier gain
3120  GOTO Pmt_stand ! restart
3130  END IF
3140  !
3150  !.............
3160  !
3170  ! low value, not yet at maximum voltage
3180  IF A<(4000*(Mpc(10)-Mpc(17))) AND Mpc(2)<253
      THEN
3190  OUTPUT @Mpc USING "#,A";"h" ! increment voltage
3200  Mpc(2)=Mpc(2)+1 ! memory of voltage
3210  GOTO Pmt_stand ! restart
3220  END IF
3230  !
```

```
3240 !.............
3250 !
3260 ! low value, at or near maximum voltage
3270 IF A<(4000*(Mpc(10)-Mpc(17))) AND Mpc(2)>=253
     THEN
3280 IF Mpc(1)<1000 THEN ! not yet at highest gain
3290 OUTPUT @Mpc USING "#,A";"R" ! turn high voltage
     off
3300 ALPHA ON ! show alphanumerics
3310 PRINT "powering down PMT to increase gain"
3320 WAIT 5 ! wait for voltage to drop
3330 OUTPUT @Mpc USING "#,A";"h" ! reset to 320 V
     minimum
3340 Mpc(2)=0 ! memory of voltage
3350 WAIT .5 ! takes a while
3360 IF Mpc(1)=10 THEN ! increase from × 10 to × 100
3370 OUTPUT @Mpc USING "#,A";"D" ! gain = × 100
3380 WAIT .1 ! takes a while
3390 Mpc(1)=100 ! memory of amplifier gain
3400 GOTO Pmt_stand ! restart
3410 END IF
3420 IF Mpc(1)=100 THEN ! increase from 100 to 1000
3430 OUTPUT @Mpc USING "#,A";"H" ! × 1000
3440 WAIT .1 ! takes a while
3450 Mpc(1)=1000 ! amplifier gain memory
3460 GOTO Pmt_stand ! restart
3470 END IF
3480 ELSE ! sometimes there just is not enough light
3490 BEEP
3500 PRINT "NOT ENOUGH LIGHT AT MAXIMUM VOLTAGE AND
     GAIN"
3510 PRINT "reduce dynamic range or adjust optics"
3520 PAUSE
3530 END IF
3540 END IF
3550 !
3560 !.............
3570 !
3580 ! high value, not yet at lowest voltage
3590 IF (A>(4000*(Mpc(10)+Mpc(17))) OR A>4000) AND
     Mpc(2)>1 THEN
3600 OUTPUT @Mpc USING "#,A";"l" ! decrement the
     voltage
3610 Mpc(2)=Mpc(2)-1 ! reset high voltage memory
3620 GOTO Pmt_stand ! restart
3630 END IF
3640 !
3650 !.............
3660 !
3670 ! high value, at lowest voltage
```

```
3680 IF (A>(4000*(Mpc(10)+Mpc(17))) OR A>4000) AND
     Mpc(2)<=1 THEN
3690 IF Mpc(1)=1000 THEN ! reduce gain from × 1000
     to × 100
3700 OUTPUT @Mpc USING "#,A";"D" ! gain = × 100
3710 Mpc(1)=100 ! memory of gain
3720 WAIT .1 ! takes a while
3730 GOTO Pmt_stand ! restart
3740 END IF
3750 IF Mpc(1)=100 THEN ! reduce gain from × 100 to
     × 10
3760 OUTPUT @Mpc USING "#,A";"B" ! × 10
3770 Mpc(1)=10 ! memory of gain
3780 WAIT .1 ! takes a while
3790 GOTO Pmt_stand ! restart
3800 END IF
3810 END IF
3820 !
3830 !.............
3840 !
3850 ! acceptable PMT output if reached here
3860 GOSUB Stabil ! wait until output is stable
3870 A=0
3880 IF Mpc(23)=0 THEN ! low precision
3890 Reps=Mpc(3)*10
3900 ELSE ! high precision
3910 Reps=Mpc(3)*100
3920 ND IF
3930 FOR I=1 TO Reps ! for each of the required
     number
3940   OUTPUT @Mpc USING "#,A";"!" ! start ADC
3950 ENTER @Mpc USING "#,W";B ! get result
3960 A=A+B ! add it into stash
3970 NEXT I
3980 A=A/Reps ! find average
3990 IF Bufsiz=0 THEN GOSUB Framit ! need new frame?
4000 X(Tally)=320+(Mpc(2)*3.5) ! update x-axis
4010 Y(Tally)=A ! update y-axis
4020 GOSUB Plotit ! plot new PMT response against
     voltage
4030 IF A<(4000*(Mpc(10)-Mpc(17))) OR
     A>(4000*(Mpc(10)+Mpc(17))) THEN ! problem
     occurred
4040 ALPHA ON ! alert operator to this rare event
4050 IF A<(4000*(Mpc(10)-Mpc(17))) THEN PRINT "GOT
     ACCEPTED THEN UNDERSHOT"
4060 IF A>(4000*(Mpc(10)+Mpc(17))) THEN PRINT "GOT
     ACCEPTED THEN OVERSHOT"
4070 GOTO Pmt_stand ! restart
4080 END IF
```

```
4090  Mpc(8)=A ! PMT output at standardization
4100  IF Mpc(20)=1 THEN CALL
      Stash_fig(X(*),Y(*),Dotter(*), (Tally-1),Leg$(*))
      ! if test, can save graphics
4110  CALL Shutter(Mpc(*),2) ! set shutters to
      dark-field state
4120  WAIT .2 ! wait in case a slow shutter
4130  GOSUB Stabil ! wait for signal to stabilize
4140  A=0 ! empty register
4150  FOR I=1 TO Reps ! for each replicate
4160  OUTPUT @Mpc USING "#,A";"!" ! start ADC
4170  ENTER @Mpc USING "#,W";B ! enter result
4180  A=A+B ! add to stash
4190  NEXT I
4200  Mpc(9)=A/Reps ! average
4210  CALL Shutter(Mpc(*),1) ! shutters back to
      measure
4220  RETURN
4230  !
4240  !........................
4250  !
4260  Measure_pmt: ! get new PMT measurement
4270  !
4280  CALL Shutter(Mpc(*),1) ! get shutters to
      measuring position
4290  IF Mpc(23)=1 THEN GOSUB Stabil ! high precision
      flag set
4300  A=0 ! empty register
4310  FOR I=1 TO Mpc(3) ! for each replicate
4320  OUTPUT @Mpc USING "#,A";"!" ! start ADC
4330  ENTER @Mpc USING "#,W";B ! get result
4340  A=A+B ! add to stash
4350  NEXT I
4360  Mpc(11)=A/Mpc(3) ! take average and put in
      answer box
4370  RETURN
4380  !
4390  !.....................
4400  !
4410  Dark: ! measure dark field (NL = no light)
4420  CALL Shutter(Mpc(*),2) ! get to dark-field
      condition
4430  IF Mpc(23)=1 THEN ! high precision flag set
4440  WAIT .2 ! to settle
4450  GOSUB Stabil ! wait for stable signal
4460  END IF
4470  A=0 ! clear register
4480  FOR I=1 TO Mpc(3) ! for each replicate
4490  OUTPUT @Mpc USING "#,A";"!" ! start ADC
4500  ENTER @Mpc USING "#,W";B ! get result
```

```
4510 A=A+B ! add into stash
4520 NEXT I
4530 Mpc(7)=A/Mpc(3) ! put average in answer box
4540 CALL Shutter(Mpc(*),1) ! put shutters back to
     measure
4550 RETURN ! finished measuring dark field
4560 !
4570 !........................
4580 !
4590 Reset: ! reset voltage and gain to targets
4600 !
4610 IF Mpc(1)<>Mpc(13) THEN ! if not already at
     target
4620 IF Mpc(13)=10 THEN OUTPUT @Mpc USING "#,A";"B"
     ! × 10
4630 IF Mpc(13)=100 THEN OUTPUT @Mpc USING "#,A";"D"
     ! × 100
4640 IF Mpc(13)=1000 THEN OUTPUT @Mpc USING "#,A";"H"
     ! × 1000
4650 Mpc(1)=Mpc(13) ! memory of current gain
4660 WAIT .1 ! why rush?
4670 END IF
4680 !
4690 IF Mpc(2)<>Mpc(12) THEN ! if not at target
4700 IF Mpc(2)<Mpc(12) THEN ! need to increase
     voltage
4710 OUTPUT @Mpc USING "#,A";"h"
4720 Mpc(2)=Mpc(2)+1 ! memory update
4730 GOTO Reset ! may need to do this again
4740 END IF
4750 IF Mpc(2)>Mpc(12) THEN ! need to decrease
     voltage
4760 OUTPUT @Mpc USING "#,A";"l"
4770 Mpc(2)=Mpc(2)-1 ! memory update
4780 GOTO Reset ! may need to do this again
4790 END IF
4800 END IF
4810 Mpc(6)=Mpc(6)-10! system has been reset
4820 RETURN ! finished resetting PMT
```

Many different types of software routines for stabilization are possible. The one that follows uses a declared escape level. Fast, approximate measurements sometimes are as useful as lengthy precise measurements. For example, if a sample is heterogeneous in nature, a large number of dispersed approximate measurements may be preferable to a few accurate measurements. This can be adjusted by setting an escape level in the subroutine for signal integration. Replication is the number of ADC cycles made for each photometer measurement and gives another method of dealing with a noisy signal. Unlike the escape level, which is determined by the nature of the experiment in progress, the number of replications is best determined

by the type of photometer in use. As well as escape levels and replication of ADC
cycles, PMT noise also may be ameliorated by damping the analog signal. Minimum
damping is used when a fast PMT response is wanted, such as when searching for
features using a scanning stage.

```
4860  Stabil: ! very simple stabilization routine
4870  !
4880  FOR I=0 TO Mpc(3)*10 ! replicates × 10
4890  OUTPUT @Mpc USING "#,A";"!" ! start ADC
4900  ENTER @Mpc USING "#,W";One(0,I) ! get result
4910  NEXT I
4920  Ave1=(SUM(One))/(I-1) ! average
4930  FOR I=0 TO Mpc(3)*10 ! replicates × 10
4940  OUTPUT @Mpc USING "#,A";"!" ! start ADC
4950  ENTER @Mpc USING "#,W";One(0,I) ! get result
4960  NEXT I
4970  Ave2=(SUM(One))/(I-1) ! make second average
4980  Esc=ABS((Ave1/100)*Mpc(16)) ! escape level
4990  IF Ave2>(Ave1+Esc) OR Ave2<(Ave1-Esc) THEN ! no
      escape yet
5000  PRINT "waiting to stabilize within";Mpc(16);"%"
5010  GOTO Stabil ! try again
5020  END IF
5030  RETURN
5040  !
5050  Pmt_end:IF Mpc(20)=1 THEN DISP "" ! clear off
      test messages
5060  SUBEND ! -----------------------------------
```

4.6   Monitoring Performance

4.6.1   Different Controllers

The CAM layout used in Chapter 3 to monitor shutter responses (Figure 3.2) also
may be used to compare the performance of different PMT controllers. As well as
controlling a Zeiss MPC 64 interface in the conventional manner over the GPIB, a
custom-built interface to a Zeiss Zonax controller also was used. The Zonax controller (originally a stand-alone controller based on an Intecolor 3650 PC) was converted
to a slave system responding to TTL (transistor-transistor logic) in spare sensor
circuits at address 23565D. The PMT was switched manually between the two
otherwise independent microscope controllers (Zonax and MPC 64), which provided
all the normal functions of a computer-assisted light microscope. Thus, different
ways of powering-up the PMT could be monitored from the parallel input to armed
multiprogrammer cards. The reason for being interested in the Zonax is that the
register of its digital transformer tap can be accessed directly by POKE commands,
allowing instantaneous programming of the PMT high voltage.

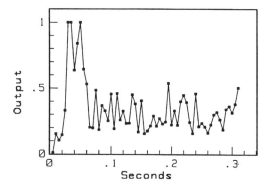

FIGURE 4.3

PMT output when powered-up with a POKE command from the Zeiss Zonax controller and monitored from a multiprogrammer ADC.

4.6.2   Transient Events

A worst-case scenario is shown in Figure 4.3, where a PMT was powered up instantaneously with a POKE command from the Zeiss Zonax controller. There was a transient saturation (output = 1) when first activated, followed by a very noisy signal, although with appropriate damping and integration it was possible to obtain accurate and reproducible measurements. An on-screen window functioning as a chart recorder (as in Figure 4.3 created by lines 2310 to 2430 in the previously listed SUB Pmt) was far more revealing than an actual analog meter, where inertia of the armature and needle integrated signals such as those shown in Figure 4.3 so that the needle was steady.

Another power-up transient is shown in Figure 4.4, 1, where an otherwise stable PMT operated in an incremental mode (incrementing the voltage in a stepwise manner) from the Zeiss MPC controller was powered-up without an adequate time (5 sec) to equilibrate to a power-down instruction (turn voltage off). The cause of the transient at around 360 V was not identified (because it was easy to prevent), but it could have originated either from the PMT (e.g., a change in photoelectric threshold or work function) or from the amplifier (e.g., a capacitive discharge). The full height of the acceptance window on each side of the target for the fraction of the dynamic range is shown by the horizontal lines in Figure 4.4 (as created by lines 1880 to 2290 in the previously listed SUB Pmt). Note how the PMT output continued to rise after incrementation of the voltage was completed (arrow in Figure 4.4, 2). Had this taken the output above the acceptance window, the voltage would have been decremented to get the endpoint back into the window by checking the drift. Thus, if a narrow acceptance window is specified and the window is repeatedly overshot or undershot, it would be necessary to change the escape level, replication, or damping to produce a slower change in the PMT output. In Figure 4.4, had the target fraction of the dynamic range been low, the transient at around 360 V would have caused premature termination of the power-up operation, leaving the PMT with a minuscule fraction

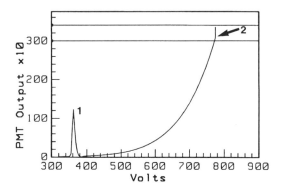

FIGURE 4.4
Power-up transient (1) in a recently active PMT, monitored by the main controller from the MPC 64. The arrow (2) indicates where output continued to rise after the last increment to PMT voltage.

of its dynamic range in operation. Allowing the operator to watch the power-up procedure with on-screen graphics is an excellent safeguard against the problem caused by output transients, although, if they are common, it would be necessary to search for them programmatically.

4.6.3   Amplifier Gain

A simple algorithm to set the output of a PMT is to increase the voltage until an output is obtained within the acceptance window, starting at the lowest amplifier gain and incrementing the amplifier if the acceptance window cannot be attained. This is shown in Figure 4.5, for a very low light intensity, where lines 1, 2, and 3 show amplifier gains of ×10, ×100, and ×1000, respectively. The right-side edge of the acceptance window in Figure 4.5 was set by the maximum voltage attainable (≈1210 V) so that, had the light intensity been any less, it would not have been possible to achieve an acceptable output (which would have prompted lowering of dynamic range fraction).

   A simple algorithm, such as the one that generated Figure 4.4, is slow and sometimes results in an inappropriate endpoint (maximum voltage at the lowest amplifier gain). The problem can be corrected by precociously increasing the amplifier gain (prompted by the occurrence of negligible PMT outputs at low voltages, as in lines 3030 to 3130 in the previously listed SUB Pmt), and by increasing the step size of the voltage increment until the output approaches the acceptance window.

4.6.4   Good and Bad PMT Settings

Figure 4.6, 1, shows a stack of absorbance profiles from a raster scan with a scanning stage. The bump in the middle was the nucleus of a muscle fiber reacted by the Feulgen

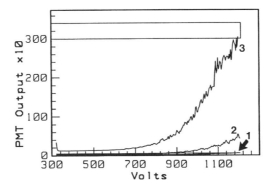

FIGURE 4.5
The result of a simple algorithm for setting a PMT. Lines 1 to 3, respectively, show outputs with amplifier gains of ×10, 100, and 1000. Results were obtained with default settings: fraction of dynamic range = 0.8, acceptance window = 0.05, escape level = 2%, replication = 20, and damping = 10 Hz.

reaction for DNA. Other possible methods for DNA include gallocyanin-chromalum after RNase, methylene blue after RNase, and methyl green (Cowden and Curtis, 1975). The operating conditions for the PMT were bad, deliberately simulating the conditions that might have been attained by a simple power-up loop (dynamic range

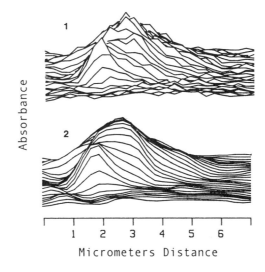

FIGURE 4.6
Absorbance profiles of a muscle fiber nucleus stained by the Feulgen reaction for DNA (0.25-μm scanning increment, ×100-NA 1.25 objective, 0.08-mm aperture). In set 1, measurements were made with a noisy, undamped PMT signal as attained by a simple power-up loop (dynamic range fraction = 1, escape level = 25%, replication = 1, and damping = 5 kHz). The absorbance maximum = 1.16. In set 2 (absorbance maximum = 1.01), made over the same grid area as the top set, the PMT was powered up with the software outlined in this chapter, using a dynamic range fraction = 0.8, escape level = 2.5%, replication = 50, and damping = 10 Hz.

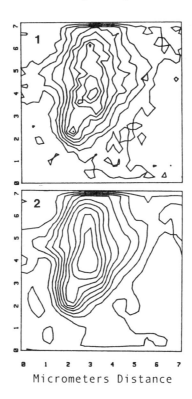

FIGURE 4.7

Contour maps of the absorbance data in Figure 4.6, with contour intervals set at one tenth of peak
absorbance.

fraction = 1, escape level = 25%, replication = 1, and damping = 5 kHz). The corresponding contour lines (Figure 4.7, 1) were almost worthless in characterizing the
appearance of the nucleus, with respect to chromatin distribution, as in the separation of mitotic satellite cells from non-mitotic nuclei incorporated into the muscle
fiber (Ontell, 1974). At more favorable operating conditions established programmatically to produce a steady signal (dynamic range fraction = 0.8, escape level =
2.5%, replication = 50, and damping = 10 Hz), the raw data (Figure 4.6, 2) required
no filtering to produce a useful contour map (Figure 4.7, 2). The pattern of contours
adequately described the subjectively homogeneous chromatin distribution, while
integration of the areas under the absorbance curves gave a measure (uncalibrated)
of DNA content.

The SUB Pmt listed earlier was developed for Zeiss MPC hardware with a
relatively large number (255) of programmable voltages but only a small number (3)
of programmable amplifier gains. Testing of three different MPC controllers showed
that signal noise was not much different at high voltage, low gain vs. low voltage,
high gain. Other hardware might differ in this respect, calling for software to minimize
the high voltage and maximize the gain (if the PMT voltage is noisier than the amplifier).

5

Monochromators and Spectrophotometry

5.1 Introduction

Chapter 4 considered the principles of photometry, but each measurement was static — one measurement, at one place, at one wavelength, in one plane of polarization, at one temperature, at one angle of stage tilt, at one pH, and at one refractive index. Now we are ready to introduce step-and-measure software, enabling the CAM to step through all these dimensions, assembling a multidimensional matrix to test a scientific hypothesis. Although this chapter deals with monochromators and spectrophotometry, the key point is that, to the CAM, wavelength is just another dimension to step-and-measure. Thus, the principles considered in this chapter are equally applicable to step-and-measure operations in other degrees of freedom. For a detailed account of monochromator optics for the CAM, consult Goldstein (1975) or Piller (1977), and, for general optical principles, Hecht (1987). Although microscope objectives are corrected for chromatic aberration, one might anticipate problems when scanning through different wavelengths, because light at different wavelengths will come into focus at different points along the optical axis of the microscope (Chapter 1). In practice, however, the problem is negligible, provided that the condenser aperture is smaller than the numerical aperture of the objective (King, 1957).

5.2 Step-and-Measure

5.2.1 Data Structure

To optimize the unlimited versatility of the CAM, the software must be interactive, with the operator choosing what is to be done, as well as performing all the numerous non-automated tasks (such as centering the target in the photometric aperture). There are two possible approaches to handling data collection and storage. An open-ended file or buffer could be created and data spooled to it, leaving the data structure to be deciphered at some later time. Alternatively, array dimensions and sizes can be specified in advance, so that ON END and ON EOT statements are used only for error trapping. Generally speaking, the latter method is worth the effort. Thus, a primary rule in programming should be that all data structures must be specified in advance of data collection, using ALLOCATE statements. However, when the CAM is used for remote measurements by fiberoptics, the depth of tissue penetration is determined manually by the operator and is not known until completion (Chapter 11). A robust solution to this dilemma is as follows.

```
[1]   ALLOCATE Leg$(5)[50]  ! always the same
[2]   Leg$(1) = "Project number and unique date-time stamp"
[3]   Leg$(2) = "Main title"
[4]   Leg$(3) = "X-axis title"
[5]   Leg$(4) = "Y-axis title"
[6]   Leg$(5) = VAL$(Total_data_number)
[7]   ALLOCATE Fig(2,Total_data_number)
[8]   FOR I = 1 to Total_data_number
[9]   Fig(0,I) = x
[10]  Fig(1,I) = y
[11]  Fig(2,I) = Degree_of_freedom ! line type for plotting
[12]  NEXT I
[13]  ALLOCATE Typ$(Total_df)[50]  ! optional
[14]  FOR I = 1 to Total_df
[15]  Typ$(I) = "Structure_code"
[16]  NEXT I
```

For example, with three degrees of freedom such as wavelength, temperature, and pH, wavelength would be loaded at [9] and Structure_code at [15] might read "pH6.5/Centigrade25", where a slash is the separator for deciphering the values for code "pH" and "Centigrade". Note that the format of Fig can handle both regular and irregular data structures but pays the price by inefficient data storage. Thus, a trio of files, such as *Leg1*, *Fig1*, and *Typ1*, is stored in a project directory. When archived, a project number may be used as a prefix (e.g., P1L1, P1F1, and P1T1) to facilitate data retrieval. *Leg* and *Typ* files can be searched and deciphered rapidly to locate or understand stored data sets.

Using OPTION BASE 0, there are a few slots remaining for special requirements. For example, line type 1 might be a mean value, and line type 2 might be a standard deviation, in which case the original number of data is stored to allow future statistical testing, Fig(1,0) = N.

5.2.2 Scanner Hierarchy

To step-and-measure through several degrees of freedom, the primary scanner should be the device having the most superior combination of speed and repeatability. For example, in measuring a series of transmittance spectra at different temperatures using a thermal stage (Chapter 12), a grating monochromator with a stepper motor is superior in speed and repeatability to the thermal stage, which may take some time to reach a temperature within an acceptance window. It is difficult, however, to establish a fixed hierarchy of scanners. Even the mechanical errors of a superior scanning monochromator are exaggerated when scanning across intense peaks such as those in the emission spectrum of a mercury arc. Thus, the best solution is to provide useful on-screen information, then let the operator decide the hierarchy of scanners. A list of options for a well-equipped CAM might be as follows.

(1) Monochromator immediately beneath a PMT for general purpose spectrophotometry

(2) Monochromator in front of a xenon arc for excitation fluorometry (Chapter 8)

(3) NIR monochromator, used when the monochromator beneath the PMT is bypassed and the PMT is replaced by a photodiode

(4) Photodiode array spectrograph (Chapter 10)

(5) Versatile actuator for polarized light analyzers and compensators (Chapter 7)

(6) Scanning stage offering two degrees of freedom (x and y position)

(7) Tilting stage (Chapter 7)

(8) Servo-motor actuator to move the distal end of an optical fiber across a remote specimen (Chapter 11)

(9) Depth detector for a hypodermic needle used for probing an optical fiber into a tissue (Chapter 11)

(10) pH controller (Chapter 12)

(11) Thermal stage (Chapter 12)

(12) Refractive index controller (Chapter 12)

5.2.3 Spectral Scanning Algorithm

A basic operation for a CAM is standardizing the spectrophotometer. Each optical component in a CAM may have its own spectral response:

(1) Emission spectrum of the light source

(2) Overall transmittance of lenses, mirrors, windows, slides, cover slips, mounting medium, and immersion oil in the light path

(3) Spectral efficiency of monochromators and their stray light filters

(4) Spectral sensitivity of the photometer

The most commonly used and reliable method is to follow the algorithm for a manual single-beam spectrophotometer.

[17]   Create an appropriate blank and find the wavelength that gives the maximum photometer response.

[18]   Adjust the photometer to give almost a maximum response.

[19]   Temporarily close the illuminator shutter to measure the photometer response (dark-field measurement).

[20]   Go to each wavelength and measure the photometer response.

[21]   At each wavelength, subtract the dark-field measurement from the photometer response measured in [20] to get a blank measurement.

[22]   Place a sample in position and go to each wavelength to measure the photometer response.

[23]   At each wavelength, subtract the dark-field measurement from the photometer response measured in step [22] to get the sample measurement.

[24]   At each wavelength, express the sample measurement as a function (e.g., transmittance or absorbance) of the blank measurement.

5.3   Monochromators

Light of one color reasonably may be called monochromatic light, and a device that eliminates all other colors to leave only one color may, thus, be called a monochromator; however, this leads to a logical trap if colors are defined by their wavelengths. For example, if a beam of monochromatic yellow light generated with an LM illuminator and prism monochromator is projected onto a screen, the light next to the short wavelength edge at 565 nm looks a bit greener than light next to the long wavelength edge at 590 nm, which looks slightly orange. A relatively wide bandpass (25 nm), such as this, will smooth over a variety of the peaks found in spectra of substances such as hemoglobin. On reducing the bandpass by narrowing the monochromator slit, the light may be made more pure, so that peaks in the spectrum can be better defined. But, pushing this procedure to its limit produces a bandpass so narrow and pure that its energy content is too low to measure. Thus, practical spectrophotometry with a CAM is a compromise between bandpass and energy level. With a strong illuminator and a large photometric aperture, a narrow (1-nm) bandpass can be used to delineate narrow peaks in a spectrum. If we attempt the same measurement with low light levels, though, we are forced to increase the bandpass to measure the spectrum, thus flattening its real, narrow peaks.

5.3.1   Software Options

The number of wavelengths to be measured in a spectrum has a direct effect on the length of time required to measure the spectrum and, if there are many measurements to be made, the operator probably will request a sensible number, such as measuring from 400 to 700 nm in steps of 10 nm. After that, the operator probably will choose a bandpass of 10 nm. To improve on the often restrictive software that comes with a commercial CAM spectrophotometer, imaginative solutions are needed for the perennial problem of balancing user-friendliness (simplicity of operation) with versatility (offering the maximum range of user choices). This is further compounded by our goal of writing software capable of surviving major changes in hardware, such as replacing a continuous interference filter with a grating monochromator. One solution is as follows:

[25]   Operator identifies the illuminator to be used from a list of known devices.

[26]   Operator identifies the monochromator to be used from a list of known devices.

[27]   Operator selects minimum, maximum, and step-size values for the spectrum.

[28]   Check that the requests are feasible. For example, a 1-nm step cannot be supported if the monochromator has a fixed 10 step, and a halogen illuminator cannot routinely be used down to 350 nm. If the requests cannot be supported, explain problem to operator, then go to [25].

[29]   Operator selects bandpass (which almost always is set manually on the monochromator).

[30]   Proceed with standardization. If successful, then SUBEXIT.

[31]   Problems have been encountered reaching the specified acceptance window for the PMT (Chapter 4). If the bandpass is low (say, <5 nm), then explain the problem and offer the operator a choice to increase the bandpass. Go to [30] if the bandpass can be increased.

[32]   Problems have been encountered reaching the specified acceptance window and the operator needs to maintain the current bandpass. Solve the problem by increasing the size of the acceptance window or lowering the fraction of the dynamic range (Chapter 4); go to [30].

"Explaining the problem" involves a multitude of helpful suggestions, such as asking the operator to check that there is a visible field to be measured and that the photometer head is passing light to the PMT. If the program fails because of impossible operating conditions, it will do so in the subprogram that operates the PMT.

5.3.2   Prism Monochromator

A 60° prism of glass effectively disperses the visible spectrum. After passing through the entrance slit, the beam is expanded to pass through the prism between the apex refracting angle and the opposite side base. Thus, the focal plane is slightly curved when monochromatic light escapes through the exit slit. The most likely application for an old prism monochromator retrieved from a basement storage area is as an excitation monochromator for fluorometry, where it may reside next to the LM. The enormous weight of many old prism monochromators discourages attempts to mount them on top of the CAM. Exposure to the immediate heat of an arc lamp, however, may soon reveal whether or not the manufacturer included a bimetallic temperature compensator, because angular dispersion through a prism is affected by temperature.

FIGURE 5.1

A grating monochromator mounted in an illumination pathway to a CAM, showing remote-local switch, 1; status diodes, 2; manual wavelength changer, 3; slit-width control, 4; the optical axis, 5; and a mounting rail, 6.

When scanning manually with a prism monochromator, it may escape notice that the relationship between prism rotation and change of wavelength is seldom linear. This provides a challenge for automation when using a knob-rotating actuator (Chapter 7). The problem may be solved theoretically if the geometrical optics of the monochromator are known, but an empirical look-up table may be just as effective. Controlling the bandpass on a prism monochromator also is difficult. Even with a constant slit width, the bandpass may change as a function of wavelength. In short, prism monochromators do not lend themselves readily to automation in a CAM but make excellent exciter filters for fluorometry, where they can be changed manually.

5.3.3   Grating Monochromator

Diffraction grating monochromators are the mainstay of CAM spectrophotometry at present (Figure 5.1). Originally, diffraction gratings were manufactured by very precise ruling engines, but now we have entered an era of mass production made possible by holographic interference diffraction gratings. A monochromator grating consists of straight, parallel, evenly spaced grooves in a layer of aluminum or gold. The optical theory of diffraction gratings is given by Piller (1977) and Hecht (1987). The main practical point that a programmer must know is that the wavelengths of

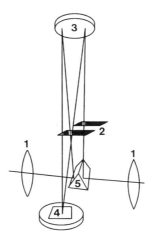

FIGURE 5.2

Components of a Zeiss grating monochromator: ultraviolet achromats, 1; adjustable slit for bandpass, 2; concave collimating mirror, 3; diffraction grating, 4; and retractable roof prism, 5.

light exiting from a grating monochromator are a function of the angle of the grating and the type of grating.

Figure 5.2 shows the layout of parts in a Zeiss grating monochromator. The monochromator is bypassed when not needed by retracting the roof prism (Figure 5.2, 5). The program must inform the operator of this possibility to have the position of the roof prism checked before attempting spectrophotometry. The angle of the grating is programmed to give the required output, usually with a stepper motor as the actuator. The maximum spectral efficiency is given by the blaze wavelength specified for the grating, which relates to the spacing and cutting angle of the grooves (the grooves are triangular in cross-section). Figure 5.3 shows smoothed curves for the grating efficiencies of the two gratings used in the monochromator of Figure 5.2.

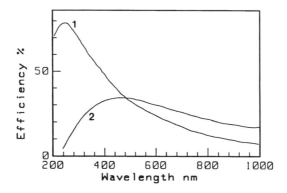

FIGURE 5.3

Smoothed curves for efficiencies of gratings for ultraviolet (1) and visible light (2).

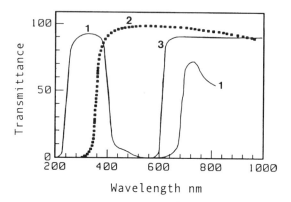

FIGURE 5.4

Stray-light filters for a grating monochromator, showing filters suitable for ultraviolet (1), visible (2), and near-infrared (3).

One grating (Figure 5.3, 1) is for UV absorbance measurements or excitation, while the other is for general use from 300 to 1000 nm. Ultraviolet absorbance measurements are particularly useful in botanical studies for the identification of aromatic compounds in cell walls (Ames et al., 1992). Smoothing is necessary for grating efficiency curves because they may show numerous irregularities, some of which, such as Wood's anomalies, may be polarization dependent. A depolarizer may be needed at an appropriate point in the light path if spectral measurements are to be made of polarized light. Also, there may be polarized light interactions between beam splitters and grating monochromators.

5.3.4   Stray-Light Filters

Another key point for the programmer is that a grating monochromator does not produce a single spectral peak output, but rather a series of harmonic peaks. For example, with a blaze wavelength of 300 nm, a second order peak also occurs at 600 nm, and a third order peak at 900 nm. Thus, a grating monochromator on a CAM should always be accompanied by a series of filters to remove the higher order harmonics, which usually are regarded as stray light. Figure 5.4 shows the stray light filters for the grating shown in Figure 5.2, 2. When assuming control of CAM programming, it is absolutely essential to check these stray light filters and match them to the grating, because the programmer must decide at which wavelength to swing in a particular stray light filter. Thus, ascending in wavelength from 200 to 1000 nm, the change from filter 1 to 2 would be placed before 400 nm, probably at 380 nm. The change from 2 to 3 at 700 nm is less critical. The curves in Figure 5.4 are only approximate, and exact bandpass spectra must be obtained from the manufacturer's documentation or by direct measurement.

Common sense problems are just as important as technical problems in programming for spectrophotometry. In the case of monochromators, the programmer must check that the correct grating is in place, and that the nominal value for the output wavelength is correct. Possibly, a grating may have been changed by a previous user. Even worse, the new grating may need a multiplier for the nominal output. Thus, step one is to use a low-power helium-neon laser or a calibrated narrow-pass filter to check the monochromator. Step two is to examine the photometer response to check that the change-over wavelength is not right on the edge of the stray-light filter bandpass. Using the extreme edge of the bandpass must be avoided, because it greatly exaggerates mechanical errors in the monochromator actuator. For example, ± 1 nm will scarcely affect the photometer response in the middle of the bandpass, but on the steep edge of the bandpass it will produce a major effect.

The swing-in time for stray-light filters also should be checked and an appropriate WAIT statement written in the software. If there is no WAIT statement, the trailing edge of the previous filter support or the leading edge of the new filter support may block light during the earliest part of the photometer integration time, thus reducing the mean value. The effects of gravity may become detectable. If a stray-light filter changer is mounted horizontally in the illumination pathway, with the mechanical pivot either above or below the optical axis, the swing time is symmetrical. But if the pivot is to one side of the optical axis, the filter will have to swing against gravity in one direction and with gravity in the other.

5.3.5   Monochromator Location

If only a single monochromator is available, there are two locations in which it may be placed. If placed in the illumination pathway, it protects sensitive samples from heat and photobleaching. This also simplifies adjustments: with Köhler illumination, the grating image should be in the sample plane, and the slit width should match the photometric aperture (Piller, 1977). Alternatively, with the monochromator mounted immediately beneath the photometer head, the sample can be viewed normally with white light, without constantly taking the monochromator off-line. Taking the monochromator off-line is something to be avoided. If the measuring aperture happens to be open, and if the PMT happens to be at a high voltage to measure monochromatic light with a narrow bandpass, then taking the monochromator off-line and admitting intense white light will saturate the PMT.

Having the monochromator almost permanently in-line below the photometer head is definitely the safest configuration. Another advantage of this location is that the effect of ambient light is reduced, because it is added to the already intense light of the illuminator; whereas, if the monochromator is located in the illumination pathway, ambient light added to weak monochromatic light at a narrow bandpass may have a large effect on the signal. It is more difficult to align a monochromator above the CAM, however, because the intensity of illumination is much lower. The room must be dark, and the operator must stand on the bench and look down the optical axis, having first removed the photometer head. The lamp image is adjusted to be in the top entrance slit, using thin paper to find the image of the filament or arc.

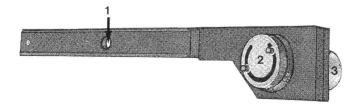

FIGURE 5.5
A Zeiss continuous interference filter, showing the optical axis, 1; manual control and limit stops, 2; and DC motor, 3.

5.3.6   Continuous Interference Filter

A continuous interference filter commonly is called a wedge, because of its structure (Figure 5.5). Extremely thin, alternating layers of strong-reflecting and weak-reflecting materials create interference effects so that only monochromatic light is transmitted. If an elongated wedge of interference layers is created, thin at one end and thick at the other, then the wavelength of transmitted light changes along the wedge. Since the whole structure is thin ($\approx$1 mm), it can be inserted directly across the optical axis of the microscope, without the need for Telan lenses to correct for increased tube length if this is fixed at 160 nm. For CAM measurements, the bandpass is determined by the photometric aperture, although, if the wedge is inserted across the illumination pathway, the observer may see a small spectral shift across a low-magnification field of view. The wedge is usually installed across the light path with a very slight tilt, to reduce glare. The continuous interference filter has some very useful properties. It is neutral to polarized light, and it has a fairly even transmittance (typically 20 to 25%) across it range, which is usually 400 to 800 nm.

5.3.7   Software

So far in this chapter we have seen that computer control of prism monochromators is best avoided and that the dominant CAM monochromator is a grating turned by a stepper motor. Deferring any further consideration of stepper motors until Chapter 6, it remains to be shown how a continuous interference filter monochromator might be programmed. Programming stepper motor movements is usually very simple, because all the work has been done by the hardware manufacturer. But, for a wedge with a DC motor, it is a little more interesting.

The basic instruction set for the simple program given below is as follows. The control address is 23564. To activate the DC motor in a direction from 700 to 400 nm, a value of X from 0 (maximum velocity) to 127 (minimum velocity) is sent to the control address with a POKE command. The motor is turned off if X = 128. With X from 129 (minimum velocity) to 255 (maximum velocity), the motor runs from 400 to 700 nm. There are limit switches at each end of the travel, but these are <400 nm and >700 nm. The status of the control address revealed by a PEEK statement is the sum of the following possibilities:

1	=	Upper limit switch not closed
2	=	Upper limit switch closed
3	=	Lower limit switch closed
8	=	Lower limit switch not closed
128	=	At a 10-nm marker

The 10-nm marker is generated by a notched disk rotating between two photodiodes (emitter and receiver). The strategy is to balance speed against precision. It is easy to stop at each 10-nm marker if scanning at a slow velocity, but scanning the whole spectrum at a slow velocity takes too long. Many of the markers may be missed scanning at maximum velocity with a slow controller. In the subprogram listed below, the motor accelerates to overcome inertia, then crawls forward to find the marker. The software notch mentioned in the remarks below is an integration window that matches the physical notch in the wavelength marker. If just a single PEEK is used, the motors stops at the start of the physical notch, not the middle. This was tried in an earlier version of the program, using a constant high velocity with a single PEEK and using inertia to carry the edge past the leading edge of the notch, but the actual stopping point was unreliable, and there was a hysteresis effect, depending on the direction of travel. In other words, the monochromator stopped at the low-wavelength side of the notch while ascending in wavelength and on the high-wavelength side when descending.

```
100    PRINT "SLAVE PROGRAM SCANS SPECTRUM FOR
       CONTROLLER"
110    PRINT "ASSUMING 10 NM STEPS AND THAT CONTROLLER
       KNOWS RANGE"
120    INPUT "ENTER MINIMUM WAVELENGTH ";MN :! USER
       INPUT
130    INPUT "ENTER MAXIMUM WAVELENGTH ";MX :! USER
       INPUT
140    GOSUB 290 : ! INITIALIZE WEDGE, GET TO 400 NM
150    DE=MN : ! DESTINATION IS MINIMUM WAVELENGTH
160    GOSUB 830 : ! GET TO DESTINATION
170    FOR HP=1 TO ((MX-MN)/10) : ! FOR EACH STEP
       WANTED BY HP
180    GOSUB 230 : ! WAIT FOR HARD-WIRED HANDSHAKE TO
       PROCEED
190    GOSUB 440 : ! INCREMENT 10 NM
200    NEXT HP : ! LOOP BACK UNTIL END OF SPECTRUM
210    GOSUB 230 : ! WAIT FOR HARD-WIRED HANDSHAKE TO
       RE-START
220    GOTO 140 : ! START NEW SPECTRUM
230    ! WAIT  <<<<<<<<<<<<<<<<<<<<<<<<<<<<<<<<<<<<<<<<<
```

```
240    POKE 23565,1 : ! TURN ON READY SIGNAL TO HP
250    IN=PEEK(23565) : ! EXAMINE STATUS
260    IF IN=1 THEN 250 : ! NOT YET READY
270    POKE 23565,0 : ! TURN OFF READY SIGNAL TO HP
280    RETURN : ! OK TO PROCEED
290    ! GET TO 400 <<<<<<<<<<<<<<<<<<<<<<<<<<<<<<<<<<
300    MO=23564 : ! ASSIGN CONTROLLER ADDRESS TO
       MNEMONIC MO
310    IF PEEK(MO)=5 THEN 340 : ! AT LOWER LIMIT
       SWITCH
320    POKE MO,0 : ! MAXIMUM VELOCITY TOWARDS 400 NM
330    GOTO 310 : ! GO BACK AND CHECK LIMIT SWITCHES
       AGAIN
340    POKE MO,128 : ! STOP MOTOR
350    POKE MO,200 : ! MOVE AT MEDIUM VELOCITY TOWARDS
       700 NM
360    IF PEEK(MO)=5 THEN 350 : ! LOOP BACK, KEEP
       MOVING
370    IF PEEK(MO)=9 THEN 350 : ! LOOP BACK, KEEP
       MOVING
380    POKE MO,128 : ! STOP MOTOR
390    MA=0 : ! ZERO MARKER COUNTER
400    PRINT "-----------------" : ! BIG DISPLAY TO
       SEE FROM DISTANCE
410    PRINT "WAVELENGTH = 400"
420    PRINT "-----------------"
430    RETURN : ! BACK TO MAIN PROGRAM
440    ! INCREMENT 10 NM<<<<<<<<<<<<<<<<<<<<<<<<<<<<<
450    IF MA=30 THEN PRINT "ALREADY AT 700 NM"
460    IF MA=30 THEN RETURN
470    POKE MO,220 : ! SET HIGH VELOCITY TOWARDS 700 NM
480    B=0 : ! START COUNTER FOR AN INTEGRATION WINDOW
490    FOR I=1 TO 10
500    B=B+PEEK(MO) : ! SOFTWARE NOTCH
510    NEXT I
520    IF B>90 THEN 480 : ! NOT YET OUT OF LAST NOTCH
530    B=0 : ! START COUNTER FOR ANOTHER INTEGRATION
       WINDOW
540    POKE MO,160 : ! REDUCE VELOCITY BUT STILL
       TOWARDS 700 NM
550    FOR I=1 TO 10
560    B=B+PEEK(MO) : ! SOFTWARE NOTCH
570    NEXT I
580    IF B=90 THEN 530 : ! LOOP BACK, KEEP MOVING
590    POKE MO,128 : ! STOP MOTOR
600    MA=MA+1 : ! INCREMENT MARKER
610    PRINT "-----------------"
620    PRINT "WAVELENGTH =";400+(MA*10);" NM"
630    RETURN : ! BACK TO MAIN PROGRAM
```

```
640    ! DECREMENT 10 NM<<<<<<<<<<<<<<<<<<<<<<<<<<<<<<<<
650    IF MA=0 THEN PRINT "ALREADY AT 400 NM"
660    IF MA=0 THEN RETURN
670    POKE MO,60 : ! MEDIUM-HIGH VELOCITY TOWARDS 400
       NM
680    B=0
690    FOR I=1 TO 10
700    B=B+PEEK(MO) : ! SOFTWARE NOTCH
710    NEXT I
720    IF B>90 THEN 680 : ! LOOP BACK, KEEP MOVING
730    POKE MO,100 : ! SLOW TOWARDS 400 NM
740    B=0
750    FOR I=1 TO 10
760    B=B+PEEK(MO) : ! SOFTWARE NOTCH
770    NEXT I
780    IF B=90 THEN 740 : ! LOOP BACK, KEEP MOVING
790    POKE MO,128 : ! STOP
800    MA=MA-1 : ! DECREMENT MARKER
810    PRINT "WAVELENGTH =";400+(MA*10);" NM"
820    RETURN : ! BACK TO MAIN PROGRAM
830    ! GET TO <<<<<<<<<<<<<<<<<<<<<<<<<<<<<<<<<<<<<<<<
840    IF DE<400 OR DE>700 THEN PRINT "OUT OF RANGE"
850    IF DE<400 OR DE>700 THEN END
860    DE=(DE-400)/10 : ! CONVERT NM TO MARKER
       COUNTING
870    IF DE=MA THEN RETURN : ! FINISHED
880    IF DE>MA THEN GOSUB 440 : ! INCREMENT BY 10 NM
890    IF DE<MA THEN GOSUB 640 : ! DECREMENT BY 10 NM
900    GOTO 870 : ! LOOP BACK UNTIL AT REQUIRED
       WAVELENGTH
```

Values of X for motor speed were adjusted empirically to give completely reliable results relative to the size of the software notch. The program, originally written for an 8-bit controller in 1981, is not transferable because of this implicit timing logic, but it is easy to see at a glance how it runs. In general principles, however, some of the key features of more versatile programming are evident in this simple example. Converting wavelengths to loop counters simplifies things from a control perspective. Certainly, it would be possible to state

```
FOR I = 400 TO 700 STEP 10
```

but this assumes a straight line relationship between I and wavelength, which may not exist for a prism monochromator. Also, scanning may not be unidirectional. For example, a temperature sequence might start at 20°, go up to 50°, down to 0°, then back to 20° for a repeat measurement to detect permanent sample alteration. Thus, at each iteration of the loop, the target temperature is taken from a pre-planned sequence (Chapter 12).

From a top-down perspective, the three requirements for a monochromator actuator are, therefore:

1. Get to the scan minimum
2. Increment by the scan step size
3. Decrement by the scan step size

The minimum value, however, for a scan may be greater than the minimum for the scanner, and this must be handled internally by the subprogram that drives all the scanning devices. There are two basic possibilities that depend on the status of the program. Running for the first time of the day, the monochromator must be initialized. For a stepper motor, for example, this will require the motor to run against its limit switches or for the operator to input information via the keyboard. The latter option is preferable, because running against limit switches can lead to unexpected results if the device has not been checked for mechanical clearance. For example, a continuous interference filter might run into an operator-adjustable stop used by a previous operator in manual measurements for a two wavelength correction for distributional error. The ultimate protection in many devices often is supposed to be a slipping clutch, but these suffer from two problems. They may have a lubricant that hardens with age, becoming stronger than the gears the clutch is supposed to protect. Or, they may suffer from the opposite problem. Once they have slipped a few times, they keep slipping even when there is no mechanical crisis. Other devices, such as a servomotor, will already know their own positions when they are activated. After initialization, the memory of the monochromator location after its last move is available from a COM block. Depending on the device, therefore, getting to the scan minimum may involve a variety of low-level internal operations which mimic those passed into the subprogram at a high level:

1. Get to the minimum for the device
2. Increment by the minimum step for the device
3. Decrement by the minimum step for the device

First, the hierarchy of scanners should be defined, such as: (1) primary scanner is wavelength; (2) secondary scanner is fluorescence excitation wavelength; and (3) tertiary scanner is thermal stage temperature. Then the control program may use high-level commands to drive all monochromators in a hierarchical sequence.

```
[33]  CALL Mono(Command)  ! Command = tertiary scanner to
      minimum
[34]  FOR T = 1 TO N3 ! for each of N3 temperatures
[35]  CALL Mono(Command)  ! Command = secondary scanner to
      minimum
[36]  FOR E = 1 to N2 ! for each of N2 excitation wavelengths
```

```
[37]  CALL Mono(Command) ! Command = primary scanner to
      minimum
[38]  FOR W = 1 to N1 ! for each of N1 measuring wavelengths
[39]  Measure
[40]  IF W < N1 THEN increment the primary scanner
[41]  NEXT W
[42]  IF E < N2 THEN increment the secondary scanner
[43]  NEXT E
[44]  IF T < N3 THEN increment the tertiary scanner
[45]  NEXT T
```

Using this control structure, two monochromators in series both may be assigned as a primary scanner, stepping together in double precision. Then, to check the bandpasses, one monochromator could be the primary scanner, while the other could be the secondary scanner. If the goal is to maximize the flexibility of the CAM, this is a way to do it.

5.4   Distributional Error

Although a CAM is very useful for the study of clear gels in developing tissue stains and in enzyme histochemistry (Tas et al., 1980; Van Noorden and Tas, 1981), most samples are heterogeneous in nature. Microscopists may be familiar with distributional error caused by heterogeneity, but a few words of introduction are required for programmers. Imagine a rectangular cuvette filled with an evenly dispersed chromophore in solution. The photometric laws (of Bouguer, Lambert, and Beer) may be used to measure the concentration of the chromophore from the cuvette dimensions and optical absorbance. But, if the chromophore precipitates into a clump which for some reason is suspended in the middle of the light path, then the photometric laws will fail because of unabsorbed white light passing around the clump of chromophore. Yet, the same amount of chromophore is in the light path. Correcting for this distributional error, which is very common in stained preparations measured with the CAM, may be done by the two-wavelength method (Swift and Rasch, 1956; Mendelsohn, 1966; Oostveldt and Boeken, 1976). The absorbance spectrum of the stained structure is measured in a situation where distribution is homogeneous and there is minimal light scattering. Two wavelengths are chosen, such that absorbance at one wavelength is half that at the other. When measuring a heterogeneous sample, measurements at these two wavelengths then are examined for their deviation from the photometric laws and corrected accordingly (Galassi and Della Vecchia, 1988). Corrections also may be required for other deviations from the photometric laws, such as those caused by uneven sample thickness (Chikamori and Yamada, 1986).

Chapter 6

Mechanical Stages and Scanning

6.1 Introduction

Scanning stages, which move a microscope slide in programmable steps along x,y coordinates (Figure 6.1), have a curious status at present. Stepper motors have been available for many years, but it was not until digital controllers became widely available in the 1970s that stepper motors could be used easily for scanning stages. The stepper motor converts bit patterns to movement patterns. This enables a digital controller to scan a variety of different scanning patterns, such as grids and raster patterns, under computer control. In the 1980s, the price of PCs dropped substantially, while their performance and user-friendliness increased dramatically, thus reducing the overall cost and increasing the effectiveness of scanning systems. But, at about the same time, television frame-grabber boards for PCs also arrived. Video image analysis, once the domain of the electronics engineer, became available to a new generation of eager PC programmers. Almost overnight, scanning stages began to look obsolete. As in the development of television, during which pioneer electromechanical systems were overwhelmed by the cathode ray tube, mechanical scanning cannot compete with video for the majority of applications. Video can grab in milliseconds a scan that may take a mechanical stage 30 minutes.

In defense of mechanical stages, however, it is somewhat premature to condemn the technology to obsolescence. Some extremely useful techniques were developed for electromechanical scanning, especially in materials science (Fischmeister, 1968; Davis and Vastola, 1977; Davis et al., 1983). Scanning stages cannot compete in assembling large images, pixel by pixel, but, by itself, video is restricted to operator control. What if 100 high-magnification video fields are needed for each sample, and there are hundreds of samples? Thus, paradoxically, as video analysis grows in importance, it has created a demand for mechanical scanning as a fetch-and-carry

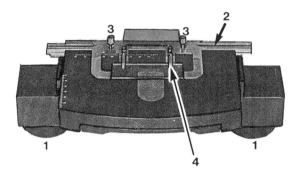

FIGURE 6.1
A scanning stage with stepper motors for *x*- and *y*-axes (1), an *x*-axis rail (2), screws to fix the slide holder in the *x*-axis rail (3), and clips to hold the slide against the stage platform (4); *y*-axis movement involves the whole stage platform.

slave (van Aspert van Erp et al., 1996; Stevenson, 1996), although this also may require automation of focusing (Kenny, 1983; Vollath, 1987) or a split-beam optical system allowing the operator to monitor the scanning operation (Tonna and Rogers, 1968).

A scanning stage still is useful for many CAM operations not requiring a massive video image. For example, an experiment may simply require the acquisition of information at a relatively small number of sites, such as scanning across a sample to obtain replicate spectra or for microelectrophoresis (Huether and Neuhoff, 1981). A scanning stage also makes an ideal optical switch for connecting optical fibers to the optical axis of the CAM. Thus, until demonstrated otherwise, it seems wise to maintain mechanical scanning as a useful accessory for the CAM.

6.2   Stepper Motors

The stator is the outer non-rotating part of the motor, while the rotor is the inner rotating part. In its simplest form, such as a four-phase stepper motor, the stator is composed of two pairs of electromagnets on teeth facing each other across the axis, and the rotor is a permanent magnet (Figure 6.2). Activation of individual electromagnets in the single-phase mode pulls the rotor into alignment, as shown in Table 6.1.

Single phase operation moves the rotor from 0°, to 90°, to 180°, to 270°, to 360°. The two-phase mode uses adjacent pairs of electromagnets, and the rotor comes to the balance point between them, pulling the rotor from 45°, to 135°, to 225°, to 315°. With pairs of electromagnets activated, the torque is increased. Half-step mode is a combination of both the single- and two-phase operations, pulling the rotor from 0° to 360° in steps of 45°. Microstepping is when the current to electromagnets is varied so that the rotor reaches balance points other than the half-way position between pairs of electromagnets. Resolution is greatly increased, but so are the cost and complexity of control circuits.

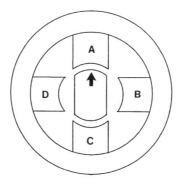

FIGURE 6.2
Four-phase stepper motor at step 0 for Table 6.1.

Instead of simply pulling a permanent magnet to an electromagnetic tooth in the stator, the variable reluctance stepper motor has opposite pairs of teeth wired in reverse to produce opposite electromagnetic polarities. The teeth of a soft-iron rotor then intersect the magnetic flux. With three phases (three opposite pairs of stator teeth), the rotor has four teeth and 12 steps per revolution. Resolution is increased by using more teeth.

6.2.1 Operational Problems

For operating scanning and tilt stages, there are design trade-offs involving gear backlash. Small movements require the motor to be geared down. When rotating in one direction, all the gear teeth are pressing in one direction but, if the motion is reversed, some of the motor rotation may be used to take up the slack between the

TABLE 6.1
Drive Modes of a Four-Phase Stepper Motor

Step	A	B	C	D	1-Phase	2-Phase	Half-Step
0	On	Off	Off	Off	Yes		Yes
1	On	On	Off	Off		Yes	Yes
2	Off	On	Off	Off	Yes		Yes
3	Off	On	On	Off		Yes	Yes
4	Off	Off	On	Off	Yes		Yes
5	Off	Off	On	On		Yes	Yes
6	Off	Off	Off	On	Yes		Yes
7	On	Off	Off	On		Yes	Yes
8	On	Off	Off	Off	Yes		Yes

teeth. Thus, for accurate placement, it may be necessary to approach the endpoint of a programmed movement from a constant direction, sometimes going past the endpoint then reversing back to it.

Control of the scanning stage may be passed to the operator for locating a small sample on a microscope slide. A slip clutch may be used for a manual override of the stepper motor, but this may lead to problems eventually. Excessive wear or loss of alignment may occur on delicate gears, or the clutch may slip so that rotor movement does not result in stage movement. This introduces one of the major problems with open-loop control systems, where there is no feedback of stage or rotor position. Thus, there is no check programmatically that an operation has been properly performed. The motor must be able to rotate rapidly for small steps at high resolution; otherwise, stage movements are too slow. If a stationary motor is instantaneously set to maximum speed, the mass of the microscope slide and friction may cause a slow start so that some of the first steps are missed. Thus, the stage may not reach the programmed end point.

These problems may be minimized by holding the microscope slide on a separate jig (rather than by sliding it over the stationary platform of the LM stage), by providing the operator with a joystick (rather than a mechanical override of the stepper motor), and by providing an automatic circuit for smooth acceleration to maximum velocity. These features should be used to full advantage if they are available. A scanning stage should be cleaned regularly to avoid small chips of glass from microscope slides increasing friction or damaging moving parts.

6.2.2   Software

A simple example of stage control is given below:

```
10    SX=23552 : ! STAGE X DIRECTION
20    SY=23553 : ! STAGE Y DIRECTION
30    FOR I=1 TO 4 : ! FOUR POSITIVE X ROTATIONS
40    POKE SX,3 : ! BIT PATTERN 0011
50    POKE SX,6 : ! BIT PATTERN 0110
60    POKE SX,12: ! BIT PATTERN 1100
70    POKE SX,9 : ! BIT PATTERN 1001
80    NEXT I
90    FOR I=1 TO 4 : ! FOUR NEGATIVE Y ROTATIONS
100   POKE SY,9 : ! BIT PATTERN 1001
110   POKE SY,12: ! BIT PATTERN 1100
120   POKE SY,6 : ! BIT PATTERN 0110
130   POKE SY,3 : ! BIT PATTERN 0011
140   NEXT I
```

A considerable amount of custom software can be developed using little more than these simple principles. PEEK statements may be used to find the status of a stepper motor. For maximum velocity, the bit patterns may be generated by an ASSEMBLER subroutine rotating an accumulator left or right.

In contrast to this relatively low-level control of a scanning stage stepper motor, high-level control may be available by working through a scanning stage controller on the HP-IB. The types of operations possible are as follows, using a Zeiss MPC controller (477464) for a fast scanning stage (471727) as the example. First, some default values are needed for velocity, distance, and direction. In normal programming, these would be determined by the operation required.

```
150    DIM Vx$[2] ! velocity on x-axis
160    Vx$="50" ! default value
170    DIM Vy$[2] ! velocity on y-axis
180    Vy$="50" ! default value
190    Dx=4 ! Dx = distance on x-axis, default
200    Dy=4 ! Dy = distance on y-axis, default
210    D1$="Y" ! default, one axis only
220    D1$="R" ! default, move right or +X
```

To block use of the joystick, the device is set to remote, with its pathway assigned to @Path.

```
230    REMOTE @Path
240    TRIGGER 7
```

Conversely, the device may be set to local to make the joystick available to the operator.

```
250    LOCAL @Path
```

A panic-button subroutine to stop a moving stage is always useful, especially if the operator can get to the keyboard before the sound of shattering glass.

```
260    REMOTE @Path
270    TRIGGER 7
280    CLEAR @Path
290    LOCAL @Path
```

Feedback from the stage is essential to find where the operator has identified the coordinates at which to start measuring.

```
300    OUTPUT @Path USING "#,2A";"X!"
310    ENTER @Path;Xko
320    PRINT "x coordinate is ";Xko,
330    OUTPUT @Path USING "#,2A";"Y!"
340    ENTER @Path;Yko
350    PRINT "y coordinate is ";Yko
```

It is possible to program velocity and distance before the stage movement is initiated. An entry of ' : : ' requests the hardware to select a safe speed to avoid missing steps. The step size is 0.25 μm.

```
360    PRINT "the current x velocity is ";Vx$
370    PRINT "enter new x velocity (2-digits or
       '::')";
380    INPUT Vx$
390    PRINT Vx$
400    PRINT "the current number of x steps is ";Dx
410    PRINT "enter new number of x steps (4-digit
       maximum)";
420    INPUT Dx
430    PRINT Dx
440    PRINT "the current y velocity is ";Vy$
450    PRINT "enter new y velocity (2-digit or '::')";
460    INPUT Vy$
470    PRINT Vy$
480    PRINT "the current number of y steps is ";Dy
490    PRINT "enter y number of steps (4 digit
       maximum)";
500    INPUT Dy
510    PRINT Dy
```

When writing new software, it is always wise to inspect the instructions being sent over the HP-IB, because most errors are created from within the new program. A formatting image is required as a line entry:

```
520    Img: IMAGE "SS",2A,4Z,"..PP",2A,4Z,".."
```

which then can be used anywhere.

```
530    PRINT "data to be sent:";
540    PRINT USING 520;Vx$,Dx,Vy$,Dy
```

The direction of scanning may be controlled as follows:

```
550    IF D1$="X" THEN PRINT "currently only one axis
       in use"
560    IF D1$="Y" THEN PRINT "currently both axes in
       use"
570    INPUT "enter X for one axis, Y for both",D1$
580    PRINT D1$
590    IF D1$="X" THEN
600    PRINT "enter R to move right    (+X)"
610    PRINT "enter G to move back     (+Y)"
620    PRINT "enter F to move forward  (-Y)"
630    PRINT "enter A to move left     (-X)",
640    INPUT D2$
650    END IF
660    IF D1$="Y" THEN
```

```
670   PRINT "enter R to move +X and +Y"
680   PRINT "enter G to move -X and +Y"
690   PRINT "enter F to move +X and -Y"
700   PRINT "enter A to move -X and -Y"
710   INPUT D2$
720   END IF
```

With everything now established, a stage movement can be triggered:

```
730   OUTPUT @Path USING 520;Vx$,Dx,Vy$,Dy
740   REMOTE @Path
750   TRIGGER 7
760   OUTPUT @Path;D1$
770   OUTPUT @Path;D2$
```

The stage coordinates can be cleared at appropriate times, such as the start of a new measuring session:

```
780   REMOTE 707
790   TRIGGER 7
800   OUTPUT @Path;"E"
```

6.3   Histochemical Mapping

From the step-and-measure approach to programming a CAM, the x- and y-axes of a scanning stage may be regarded as primary and secondary scanners, allowing a matrix of photometric measurements to be collected. To justify the use of a scanning stage, however, a more interactive approach is needed, as in the following example.

Some enzymes and substrates have characteristic intracellular patterns that can be mapped histochemically with a CAM scanning stage. For a simple raster scan, the best results are obtained from an isolated cell surrounded by clear space, such as a blood cell in a smear preparation. Unfortunately, most cells in multicellular organisms are tightly packed to form tissues and organs. While considerable progress has been made in the recognition of objects by computer (Russ, 1989), automated delineation of tightly packed cells is difficult because cell boundaries seldom are distinct. Thus, the operator may be required to delineate the cell subjectively working from the digitized image with a hand-held device; however, it is also possible to delineate areas for measurement using the scanning stage directly. The optical axis of the LM is marked by eyepiece cross-hairs, and the operator has one hand on the fine focus of the LM and the other on a small keyboard that controls the scanning stage. The area to be measured is delineated by moving the sample with respect to the optical axis, then the CAM scans within the delineated area. The abnormal accumulation of glycogen in muscle fibers is used as a demonstration.

6.3.1   Hardware

A broadband green filter (peak 555 nm) was used to measure 10-μm cryostat sections of skeletal muscle in which glycogen had been stained by the periodic acid-Schiff (PAS) reaction. Köhler illumination was obtained with a relatively large manual field aperture (about half of visual field diameter) so that much of the specimen was visible during scanning operations. This enabled the operation of the system to be checked subjectively while it was running. The photometer head was a type 01K (Zeiss 477304). The objective was a Plan ×100, NA 1.25, oil-immersion lens.

The scanning-stage keyboard consisted of four push-button transient-on switches arranged in a compass pattern (N, E, S, and W). For right-hand operation, N (finger two) moved the stage in the $+y$ direction, W (finger one) moved $-x$, E (finger three) moved $+x$, and S (finger two) moved $-y$. A fifth switch used as a signal was mounted 10 cm above switch N. Stage coordinates were shown on a solid-state display for debugging or for returning to a site of interest.

6.3.2   Software

The normal sequence of operations for working on a digitized image obtained from a television camera is to grab the whole frame and then delineate an area to be measured within the frame. However, since it takes a long time to digitize a field with a scanning stage, the normal sequence of operations may be reversed and the area to be measured is delineated first. Only the circumscribed area is scanned and no extraneous data are collected.

To avoid large raster scans that take too long to collect or which are too large for the available RAM, the operator is prompted to specify the number of scanning steps (0.25 μm) for each movement of the stage. For example, n = 1 for scanning a small part of a muscle fiber or n = 4 for a whole muscle fiber. At this point, a photometric aperture is selected with a diameter slightly smaller than the stage movement.

Arrays are dimensioned to be only slightly larger than that required for the area to be delineated. Thus, the operator is prompted to move to a point just above the top of the area to be delineated (hit signal switch), move to a point just below the area (hit signal switch), move to a point just left of the area (hit signal switch), and move to a point just right of the area (hit signal switch). This is required for the efficient storage of arrays on disk and to set the size of the frame (unused field) around graphics displays.

The operator is prompted to move to the top of the area to be delineated (bearing in mind that the measuring aperture should not move outside this area) and to hit the signal switch. The area is delineated in a clockwise direction using the direction switches of the keyboard. In most cases, it is possible to close the perimeter, but an option should be provided for automatic closure if the starting point has been forgotten (hit signal switch). Following closure of the perimeter, the area is scanned in a raster pattern within the perimeter. The time to scan a large muscle fiber (as in Figure 6.3) is about 15 minutes.

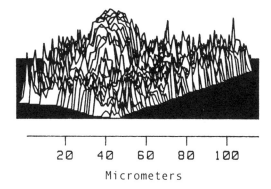

20   40   60   80   100

Micrometers

FIGURE 6.3

Three-dimensional view of absorbance in a transverse section of a muscle fiber with a large central core of glycogen (periodic acid-Schiff reaction).

The position (row and column) of each scalar (absorbance > 0) in the data array corresponds to its spatial coordinates (y- and x-axes) in the raster scan. Positions beyond the perimeter of the delineated area are empty (absorbance = 0). When the photometer is standardized on a blank area to one side of the section, absorbance is always > 0 with tissue present in the light path, even if it is unstained (as in the case of myofibrils). Thus, absorbance ≤ 0 within the delineated area indicates an error, such as damage caused by an ice crystal during freezing of the specimen or excessive noise from the photomultiplier.

Operation may be returned to the PC for the following options: (1) to store data using logical interchange format, (2) to display data three-dimensionally (absorbance on the z-axis and spatial coordinates on the x- and y-axes), (3) to display absorbance intensities as a gray map, (4) to display absorbance intensities as a contour map, and (5) to separate the scanning matrix into concentric zones.

Gray maps date from the time when line printers were used to plot pictures. A similar concept may be used with contemporary devices such as the laser printer to produce a pseudo-gray map in which the probability of a pixel being turned on is proportional to the intensity or gray level at each coordinate (dithering). Concentric zones may be defined with a simple algorithm that erodes the target area from four directions. The shape of the target area is represented in the matrix by the presence of data (absorbance > 0) and is surrounded by a frame of unused pixels (absorbance = 0).

[1]   At the start of the algorithm, an outer concentric zone is eroded from the target area (absorbance > 0) by processing the top and bottom rows of the target area plus the first and last column of each intermediate row in the target area.

[2]   These data are used to find the mean absorbance of the zone (sum of absorbance/sum of pixel areas) and then are replaced in the matrix by corresponding negative values (original absorbance × –1).

[3]   The matrix now contains a frame of unused pixels (absorbance = 0), surrounding a hollow ring defining the concentric zone (absorbance < 0), surrounding a core of

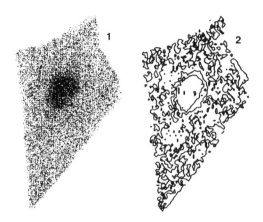

FIGURE 6.4
Gray map (1) and contour map (2) of absorbance values of a glycogen core.

uneroded data (absorbance > 1). The next concentric zone is eroded by returning to
the start of the algorithm again [1].

On completion, the original matrix may be regenerated from the negative values
(final matrix × −1).

6.3.3   Results

Isolated skeletal muscle fibers are cylindrical in shape but, when packed together,
they tend to be prismatic with polygonal transverse sections. Figure 6.3 shows a
three-dimensional view of the absorbance values corresponding to one fiber delineated from within a bundle of fibers. The step size for scanning was 1 μm, and 3846
pixels were used. Figure 6.4, 1, is a pseudo-gray map of absorbance values in this
fiber. Each scanning pixel has been expanded to a set of LaserJet pixels in which the
probability of a pixel being turned on is proportional to the absorbance in the original
pixel. Linear interpolation was used to get a smooth transition. The dark circular area
in Figure 6.4, 1, corresponds to a large core of glycogen-rich sarcoplasm. The high
degree of variability in the absorbance of the sarcoplasm surrounding the core was
caused by unstained myofibrils. Thus, absorbance was high when the measuring
aperture stopped on a relatively wide band of intermyofibrillar sarcoplasm and was
low on an unstained myofibril.

Figure 6.4, 2, is a contour map of Figure 6.4, 1, with the contour intervals at
about one quarter of the range in absorbance values. A glycogen core is clearly
visible, but little or no structure is evident in the surrounding sarcoplasm with high
variation in absorbance. The same conditions of measurement (section thickness
and the size of the scanning step and aperture) were used to remeasure the core of

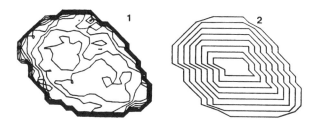

FIGURE 6.5
Contour (1) and concentric zones (2) of a glycogen core.

the fiber. Since distributional error is minimal when many measurements are made with a small aperture (Piller, 1977), the product of the mean absorbance and the area delineated gives a relative measure of the amount of PAS-stained material (essentially glycogen) in the area. By comparing the data for the whole fiber and for the core alone, it was found that 23% of the PAS-stained material was in the core of this fiber.

Skeletal muscle fibers receive their supply of oxygen from capillaries on the fiber surface. Although there are many modifying factors such as the mean radius of the fiber, the frequency of capillaries, and the concentration of myoglobin within the fiber, it is probable that oxygen concentration is greater in the outer zones of a fiber than in the axis. Thus, the activity of aerobic enzymes such as succinate dehydrogenase often is concentrated peripherally under the cell membrane, with meaningful differences between fast- and slow-contracting muscle fibers, and changes as the muscle fiber grows radially (Swatland, 1984a, 1984b, 1985a, 1985b). A comparable but opposite pattern may be found less frequently in the radial distribution of glycogen (Swatland, 1990a), the main substrate for anaerobic metabolism (Hammersen and Messmer, 1984). Figure 6.5, 1, shows a contour map of a glycogen core. The splitting of an irregularly shaped data set into isomorphic concentric zones is a lengthy procedure if absorbance values at interpolated positions must be calculated from the surrounding data. This was not attempted. Figure 6.5, 2, shows the results of the simple algorithm for unpacking data in concentric zones. The emptying of top and bottom rows plus first and last columns of intermediate rows produced a pattern that converged on a rectilinear shape whose orientation reflected x- and y-axes of the scanning stage rather than the shape of the original area that was delineated. Concentric zones were only approximately isomorphic, but their shape was not unreasonable when viewed subjectively. The mean radius of each zone decreased centripetally. Absorbance was higher in the central zone than in the outer zone. The centripetal gradient was 0.018 absorbance units per zone ($P < 0.005$; from the outer to the inner concentric zones shown in Figure 6.5, 2). As shown in the map, the steepest part of this gradient was at the edge of the core. The same technique may be used for any visible histochemical reaction product.

6.4   Switching Optical Fibers

Another application more suited to electromechanical scanning than video is the repeated measurement at specified x, y coordinates. For example, a group of cells may be treated histochemically to initiate a slow reaction resulting in a visible final reaction product. With a scanning stage, it is possible to make repeated measurements of the cells to gather information on reaction rates (Nolte and Pette, 1972). The same logic, identifying a set of coordinates then repeatedly returning to them, allows a set of optical fibers to be used for remote measurements (Chapter 11).

The x, y coordinates of the centers of optical fibers are identified by the operator, using low magnification and with the optical fibers illuminated from their distal ends. With the scanning stage at the first fiber, the magnification is increased. Usually this de-centers the first optical fiber, which then must be re-centered at high magnification. A photometric aperture is selected slightly less than the diameter of the fiber. Each fiber must be treated separately for standardization, but this merely requires the fast stage to be designated as the primary scanner and, for repeated measurements, time as the secondary scanner. Thus, a matrix of coordinates is maintained that enables the interpretation of high-level primary scanner commands (Chapter 5). Thus, the command to set the scanner to the minimum is interpreted as to move to the x, y coordinates of the first optical fiber. The command to increment the scanner is interpreted as to move to the x, y coordinates of the next optical fiber.

6.5   Tilting Stages

Conventional LM tilting stages are usually massive devices for the precise, static orientation of mineral specimens; however, a much lighter and dynamic tilting stage may be created simply by mounting a stepper motor on the LM stage. A small platform on which the sample is to be mounted is attached directly on the shaft of the stepper motor. A reasonably high resolution for stepping is required, such as step movements of 0.45°. The top of the tilting stage should be set below the axis of rotation of the stepper motor using a crank (Figure 7.7, 9). Then shims may be used to elevate the top of the sample exactly to the axis of rotation. If parts are available, mount the whole assembly (stepper motor plus tilting stage) on a rotary, gliding stage to facilitate the centering and rotation of the sample in the field of view. Images obtained at different degrees of tilt may be combined using Fourier space techniques from X-ray crystallography to analyze the three-dimensional structure of the sample (Shaw, 1990).

Chapter 7

Polarized Light

7.1   Introduction

Polarized light is an extremely useful tool for the investigation of birefringent structures such as plant cell walls, chloroplasts, nerve fibers, muscle fibers, spermatozoa, chitin fibrils, and polysaccharide gums (Ruch, 1966; Decarvalho and Vidal, 1996; Schorsch et al., 1995). Polarization allows the LM to probe structures smaller than the wavelength of light (Arimoto and Murray, 1996). Birefringent structures exhibit two different refractive indices. Thus, a transmitted light ray splits into two components traveling at different velocities, the ordinary (o) and extraordinary (e) rays, with o ⊥ e. Birefringence is measured as the refractive index (n) of the extraordinary ray minus that of the ordinary ray, and may be positive or negative in sign.

7.1.1   Principal Axes

For microscopy, an azimuth is an angle measured relative to a north-south axis of the microscope tube. In normal use, the primary north-south axis divides the visible field into left and right sides and corresponds to a position of 0° on the first polarizer, usually below the substage condenser. Caution is needed if the orientation of the visible field has been altered by LM accessories, which is why Bennett (1950) defined the 0° axis relative to the stand of the microscope. From the 0° position, it is customary to follow the convention used in the mathematics of polar coordinates, moving counterclockwise to increment the angles, as shown in Figure 7.1. Points of the compass also are used to describe the orientations of components used for polarized LM and are abbreviated as N, S, E, and W.

Relative to specimens examined under the microscope, the α or fast axis corresponds to the direction of the minimum refractive index, the minimum dielectric

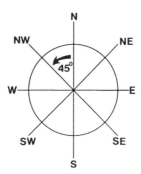

FIGURE 7.1
A widely used convention for relating angular measurements to compass points (Bennett, 1950): The zero azimuth (N) is defined by the first polarizer in the system, usually that of the substage condenser.

constant, and the maximum velocity. The γ or slow axis corresponds to the maximum refractive index, the maximum dielectric constant, and the minimum velocity. In special cases, a β axis is recognized with intermediate properties between α and γ. When working with birefringent fibers, birefringence usually is taken as positive when the γ axis is parallel to the longitudinal axis of the fiber.

7.1.2   Types of Birefringence

Intrinsic birefringence originates from molecular structure, such as an assemblage of parallel rod-like protein molecules, and is independent of the surrounding refractive index. Form birefringence originates from microstructure, such as parallel structural fibers, and does depend on the surrounding refractive index (Pimentel, 1981). Strain birefringence originates from external forces acting on molecular structure, as can be demonstrated rheologically in biological fibers (Chien and Chang, 1972). Flow birefringence is caused by shearing alignments in a flowing system.

7.1.3   Retardation

Retardation is a decrease in the velocity of light caused by an interaction with the medium through which the light is passing. Phase retardation is an interference caused by ordinary and extraordinary rays diverging and taking different paths through the specimen (one path longer than the other). When the rays recombine after passing through the specimen, they are out of phase by an amount equal to the path difference (between long and short paths). Thus, the path difference depends on both the degree of birefringence (divergence of paths) and the thickness of the specimen:

$$\text{phase retardation} = (n_e - n_o) \times \text{thickness}$$

Dichroism occurs in structures such as dyed textile fibers, the fovea of the eye, DNA, and viruses, when there are differences in the absorbance of ordinary and extraordinary rays (Oster, 1955). Methods for its detection with the polarizing microscope are described by Goldstein (1969).

7.1.4   Interference Colors

Interference colors caused by birefringence are accurately described by the Michel-Lévy color charts found in mineralogy textbooks (Hurlbut, 1959). For a subjective description, consider the colors seen on a thin slick of oil or gasoline on a puddle of water. The thinnest layer farthest from the source of the oil is white, increasing to a red-orange (from zero to first-order interference). Second-order interference occurs with progressively greater thickness of the oil slick and resembles a bright rainbow from purple through blue, green, and yellow to orange-red. Third-order interference is similar to second order but is misty and obscured. It ranges from violet, through sea-green and fleshy red, to dull purple. Consult a Michel-Lévy chart to visualize these colors properly. After measuring birefringence with new software, check the measured path difference to that on the color chart, although it may not be exact because of color rendition in printing the chart. For the thickness at which biological tissues normally are sectioned (≈ 10 μm), the birefringence of protein fibers is typically a first-order white or pale yellow. Thus, because of a high water content, tissue birefringence generally is weak. The higher order colors typical of mineralogy are only seen in thick samples, such as dissected fibers, or in mineral inclusions in tissues.

7.1.5   Types of Compensators

A polarizing microscope for transmitted light usually has a polarizer at a fixed azimuth beneath the substage condenser and an analyzer in the microscope tube above the objective. Typically, a compensator is inserted at 45° beneath the analyzer (Figure 7.2). Some compensators (such as the de Sénarmont) are fixed in azimuth, and measurements are made with a rotary analyzer with a variable azimuth. For other compensators, the analyzer is fixed and the compensator is rotated or tilted to make measurements in the axis of the microscope (Figure 7.3). Especially when measuring small path differences, objective measurements by photometry are preferable to subjective methods (Roche and Van Kavelaar, 1989).

The choice of appropriate compensator is made by reference to a Michel-Lévy chart or to a large-range compensator to find the maximum path difference. Then, a compensator of the appropriate range is used for accurate measurements, as shown in Table 7.1.

Knowing that users of the polarizing LM for mineralogy are well versed in the use of compensators, the following algorithms are for the benefit of biologists measuring weak birefringence with de Sénarmont and Brace-Köhler compensators.

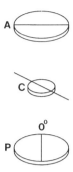

FIGURE 7.2
A typical layout for polarizer (P), compensator (C), and analyzer (A) for transmitted light. The axes of polarizers and analyzers may be checked by rotation, but not all compensators have their γ axis in line with the length of the 45° slot into which they are inserted.

7.1.6   de Sénarmont Compensator Algorithm

The use of a quarter-wave plate for the measurement of small path differences by ellipsometry was invented by de Sénarmont (1840). The original method was for light reflected from minerals, but it was adapted for transmitted light by Goranson and Adams (1933) to study strain birefringence in glass. The method is very sensitive to the molecular causes of birefringence in biological structures (Nollie et al., 1996), and the theory is described fully by Hartshorne and Stuart (1970).

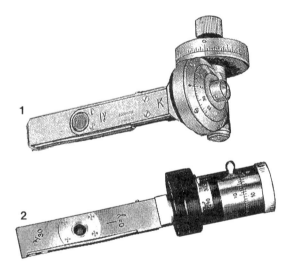

FIGURE 7.3
Ehringhaus rotary calcite compensator (1) and Brace-Köhler rotary mica compensator (2).

TABLE 7.1
Choice of a Compensator

Generic Name	Eponym	Maximum Path Difference
Calcite tilting	Ehringhaus	36.5 orders
Quartz tilting	Ehringhaus	2nd order
Mica λ / 4	de Sénarmont	1st order
Rotary λ / 4	Brace-Köhler	1/4th order
Rotary λ / 10	Brace-Köhler	1/10th order
Rotary λ / 20	Brace-Köhler	1/20th order
Rotary λ / 30	Brace-Köhler	1/30th order

The de Sénarmont compensator may be a fixed quarter-wave plate oriented diagonally in a slider that inserts into a NW-SE slot below the analyzer, so that the final orientation of its γ axis is N-S. Ideally, the manufacturer's instructions and de Sénarmont constant should be available, in which case this algorithm will not be needed. But documentation often gets lost.

[1]   Adjust Köhler illumination with a strong light source and a small field aperture.

[2]   Set polarizer E-W and analyzer N-S.

[3]   Rotate object to extinction (darkest) position.

[4]   Rotate stage 45°.

[5]   If unknown, find the γ axis of the specimen using a quartz wedge. With the γ axis of the wedge parallel to the γ axis of the specimen, the path difference is increased (additive position). With the γ axis of the wedge perpendicular to the γ axis of the specimen, the path difference is decreased (subtractive position). Remove quartz wedge.

[6]   If interference colors are greater than first-order colors, then switch to an Ehringhaus compensator.

[7]   Set monochromatic light using an interference filter, as indicated on the compensator (usually 546.1 nm or 589.3 nm).

[8]   Insert the de Sénarmont compensator into its 45° slot.

[9]   Rotate the analyzer through the compensation angle (A°) required to restore the extinction position (specimen as dark as possible against a bright background). If the γ axis of the specimen is NW-SE, the analyzer is turned negatively and, with A° < 90°, the angle to use (U°) is 90° − A°. Alternatively, with A° > 90°, U° = 90° + (360° − A°). If the γ axis is SW-NE, turn the analyzer positively starting at 90°, then U° = A° − 90°.

[10] Calculate the path difference (Γ) from the de Sénarmont constant for the particular compensator in use (for example, 3.03 nm per degree [nm/°] using monochromatic light at 546.1 nm, or 3.27 nm/° at 589.3 nm):

$$\Gamma = U° \times \text{de Sénarmont constant}$$

For a single measurement of each specimen, there is obviously no need for a CAM, although the PMT output using a small photometric aperture enables a more objective determination of the extinction position than is possible subjectively. The value of the CAM is that the extinction position can be found automatically over a long period of time, during which the sample temperature (Swatland, 1989a) or pH (Swatland, 1989b) can be changed under program control, as demonstrated later in this chapter. Also, it is possible to improve on the measurement of a single extinction point by making multiple measurements over a wide angle using a photometer (Laughlin et al., 1985). Another option is a Foster prism to separate the orthogonal components after the de Sénarmont compensator, using two photodetectors and wide-angle measurements (Lansing Taylor and Zeh, 1976).

7.1.7   Brace-Köhler Compensator Algorithm

[1]   Set polarizer W-E.

[2]   Set analyzer N-S.

[3]   Rotate the stage to find the extinction position, then orient the γ axis at 90° to the slot that holds the compensator (assuming that it is a 45° slot).

[4]   Set monochromatic light with an interference filter, as indicated by the nanometer wavelength (Wl) on the compensator, for which the Brace-Köhler constant (Bk) must be known. If Bk is unknown, but the compensator has a serial number, it may be possible to obtain a copy of the calibration sheet from the manufacturer.

[5]   Insert the Brace-Köhler compensator, rotate it through 360°, using the four extinction positions (E1 to E4) to find the path difference (PD), as follows.

$$
\begin{aligned}
A1 &= E1 * 2 \\
A2 &= 180 - (E2 * 2) \\
A3 &= (E3 * 2) - 360 \\
A4 &= 540 - (E4 * 2) \\
A &= (A1 + A2 + A3 + A4)/4 \\
PD &= (Wl/360)*ATN(2*TAN(180/Wl)*Bk)*SIN(A)
\end{aligned}
$$

7.2   Knob Rotation Actuator

Commercial CAMs are available for making polarized light measurements, such as the Leica FTM 200 used in making thin film measurements, but they are rare items outside the semiconductor industry. Most polarizing microscopes are manual and require the operator to turn analyzer or compensator knobs at various locations on the microscope, depending on the application. Thus, the first requirement for adapting a polarizing LM for computer control is an actuator to turn knobs. There are many different ways in which this might be done. The main goal of the actuator described here is not to lock out the operator, as happens if analyzer and compensator knobs

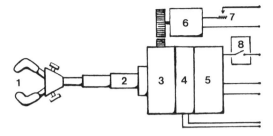

FIGURE 7.4

Actuator for rotation of analyzers and compensators: jaws to grip knob, 1; telescopic shaft, 2; variable-speed gear box, 3; electric clutch, 4; reversible motor, 5; multi-turn potentiometer, 6; adjustment potentiometer, 7; and breaker or limit switches, 8.

are replaced by dedicated stepper motors. Another point to note is that birefringence can be measured with analog electronics (Ruch, 1951; Allen et al., 1966; Taylor, 1975), thus providing a completely different approach to automation of measurements.

The actuator is mounted on a frame allowing it to approach the microscope from the main points of the compass. Adjustable jaws (Figure 7.4, 1) grip any knob and turn it. A thick rubber coating on the jaws prevents damage to the knob, as well as creating a weak link to prevent damage from programming errors during initial stages of development. In normal operation, the jaws are tightly closed to prevent slippage. The jaws are rotated by a telescopic shaft with a square cross-section (Figure 7.4, 2). This allows an analyzer or compensator to be withdrawn from the optical axis when required, without disengaging the actuator. The drive shaft operates from a variable-speed gearbox (Figure 7.4, 3), with the gear ratio matched to the degree of rotation required (fast for large angles, and slow for small angles). An electric clutch (Figure 7.4, 4) in the power train allows manual operation of the analyzer. Thus, when the clutch is disengaged and the knob is turned manually, the PC can still keep track of the knob position and resume control at any time. The motor (Figure 7.4, 5) is a fixed-speed, reversible DC motor in which the power supply is maintained through a relay (Figure 7.4, 8) dependent on a series of breaker switches. If required, the breakers may be located on the driven device to act as limit switches preventing damage from an overrun. If any one limit switch is opened, the DC motor stops and the clutch is disengaged. A multi-turn precision potentiometer (Figure 7.4, 6) gives the degree of rotation of the analyzer or compensator (Dvorak et al., 1971) with a resolution of $0.25°$. With averaging of the angle before and after an extinction position, the computation resolution is $0.125°$, which is equivalent to a path difference of 0.38 nm on a typical biological specimen measured with a de Sénarmont compensator.

7.2.1   Operational Features

(1)   The analyzer or compensator azimuth may be read directly by the operator at any time. This allows automated and operator control to be interchanged.

(2)   The main potentiometer (Figure 7.4, 6) being geared to the analyzer or compensator cannot be freely rotated, but the overall resistance can be adjusted by another potentiometer in series with the main one. When the system is initialized, the operator is required to check that the analyzer or compensator reads 0° and to adjust the resistance if it does not. This allows great flexibility if the main N-S orientation of the microscope is changed for some reason. Instead of altering all the azimuthal instructions in the software (as seen in the algorithms above), the CAM is operated with an offset (say, 90°) which is invisible to the software.

(3)   Overshoots or undershoots occurring under program control can be monitored and corrected, if necessary. This allows the actuator to be programmed for fast rotation with a wide window of acceptance, or slow rotation with a small acceptance window. The key to efficient operation is to balance speed against required resolution.

(4)   The analyzer or compensator position may be controlled and monitored as a continuous variable. This allows nonlinear control functions, such as using the actuator to operate a prism monochromator with a logarithmic or polynomial relationship between knob position and wavelength.

7.3   Continuous Measurement of Birefringence

Subjective determination of an extinction position may be fairly reliable for a static sample. But problems arise when the sample is slowly changing, because the operator may be influenced in some way by the anticipated change. An example of a continuous measuring application is given below to show the problems involved. The birefringence of striated skeletal muscle fibers has been investigated for many years (Engelmann, 1878), and it is still useful in studying muscle contraction and protein chemistry (Jones et al., 1991). Had early studies on muscle birefringence been taken seriously by biologists, the sliding filament theory of muscle contraction might have been discovered a lot sooner than it was (Huxley, 1980).

7.3.1   Hardware

A Zeiss universal microscope with polarizing attachments was operated from an MPC 64 control module over the GPIB. The main controller was an HP 9826 BASIC workstation. Custom-made components were controlled using an HP 6942A Multiprogrammer. The PMT was a side-window Hamamatsu HTV R 928 with S-20 characteristics operated from a Zeiss PMT controller, as described in Chapter 4. The illuminator was a 100-W halogen source operated from a stabilized power supply (Zeiss 487332). Köhler illumination was obtained using a solenoid-activated field aperture (Zeiss 477354). For the Zeiss compensator 473715, the de Sénarmont constant was 3.03 nm/°.

Individual striated skeletal muscle fibers were secured across a small metal frame immersed in Sørensen's phosphate buffer. The sample was maintained in a small perfusion chamber with a depth of 3.5 mm. The chamber and the apparatus to control the pH of the buffer are described in Chapter 12. Zeiss long-distance (LD) epiplan objectives were used because of the long working distance into the chamber. Most measurements were made with a ×40, NA 0.6 LD epiplan. Although not designated as a strain-free POL objective, little or no photometer change could be detected when the objective was rotated between crossed polarizers. For illumination of the sample, LD objectives were mounted on a ring (Zeiss 465545) on the substage condenser. The illumination wavelength (589 nm) was obtained from a continuous interference filter monochromator (Zeiss 474311) located above the luminous field stop. Ambient illumination was corrected when the dark-field current of the photo-multiplier was corrected, using an illumination shutter (Zeiss 467225). All measurements were made with a very low level of ambient light, as discussed in Chapter 3. The photometer head (Zeiss 477310) had a mirror with a half-silvered aperture that enabled the location of the aperture to be seen in the field of view. This was necessary because of sample movement.

7.3.2   Software

For automated measurements, the protocol was as follows.

[1]   Manually orient the muscle fiber SW-NE with Köhler illumination at 589 nm, using the photometric measuring aperture at the mid-diameter of the fiber.

[2]   Manually rotate the analyzer to the extinction position.

[3]   Automatically set the photometer using 0.3 of dynamic range.

[4]   Automatically set pH to programmed value (for example, from pH 7.0 to pH 5.0 with a step size of pH 0.5).

[5]   Automatically measure the extinction angle 20 times.

[6]   Return to step [4] until each pH has been measured.

[7]   Display and store data.

7.3.3   Results and Practical Problems

If the muscle fiber moves away from the optical axis of the microscope, this changes the depth of the sample in the field and invalidates the path difference (because path difference depends on both birefringence and sample thickness). Similarly, if the orientation of the fiber changes, then the requirements for de Sénarmont ellipsometry are not satisfied. Thus, interrupt switches located near the microscope were incorporated to allow the operator to pause the program, adjust the specimen, and return to an earlier point in the measuring loop (before measurements were invalid). Another option was to terminate the measuring loop prematurely (if a fiber became loose or obscured by an air bubble). Thus, measurements were started under operator control and full automation was only adopted for stable samples. For a completely manual

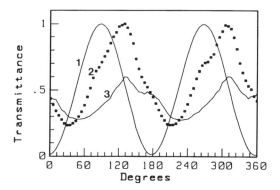

FIGURE 7.5

Transmittance as a function of analyzer rotation with a de Sénarmont compensator in the light path. Line 1 is for a blank area without birefringence, and lines 2 and 3 (standardized separately from line 1) are for a muscle fiber at pH 7.2 and 5.6, respectively.

experiment with 20 measurements at each of 14 pH values, the time required was essentially the whole working day (including data recording, calculations, and graphical presentation of results). Using computer assistance, this was reduced to about 30 minutes once the sample was in position. Scientifically, the greatest advantage of computer assistance was objective determination of the extinction position.

A number of potential problems are illustrated in Figure 7.5. Line 1 in Figure 7.5 shows the transmittance of a blank area adjacent to a muscle fiber being measured by ellipsometry. Ideally, with no birefringence in the field, the extinction position indicated by the minimum transmittance should be at exactly 0 and 180°. Corrections for mechanical imperfections and optical anomalies were made with the trimming potentiometer, adjusting the extinction positions close to 0 and 180°. In Figure 7.5, lines 2 and 3 (standardized together but independently of line 1) show a muscle fiber at pH 7.2 and 5.6, respectively. The decrease in pH caused a change in the transmittance of the fiber such that maximum transmittance was reduced while minimum transmittance was increased. This effect confounded an early algorithm that attempted to use a fixed transmittance minimum to identify subsequent extinction positions. This algorithm worked well until the pH was changed.

A more robust algorithm to find the extinction position was to move in and out of the extinction position with a rotational range of 1°. The extinction angle was taken as the angle with the minimum transmittance. If either the first or last scalar in the vector was the minimum, this indicated that the true minimum might have been beyond the range of the vector. The rotation range was then increased by 1° and the operation was repeated progressively until a true minimum was found. This extinction angle was then used as the center of the 1° rotation range for the next measurement. This algorithm was able to follow a rapidly changing extinction position without increasing the rotational range (which would have slowed the rate of measurement). The time taken to find an extinction position was from 1 to 2 seconds.

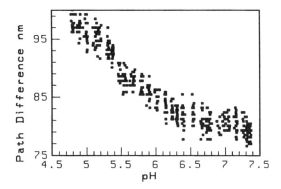

FIGURE 7.6

Relationship between optical path difference and pH in a muscle fiber at about 2 hr postmortem (ΔpH = –0.05 pH/min). Measurements and pH changes were totally automated, with the observer checking sample position and orientation.

Another problem evident in Figure 7.5 is that the muscle fiber underwent cytoskeletal disruption, allowing some fibrils to get out of register or to become kinked. Instead of describing a sine wave, the signal is asymmetrical. At a low pH (Figure 7.5, 3) the asymmetry is more conspicuous than at a high pH (Figure 7.5, 2). In muscle fibers with orderly sarcomeres in their original arrangement, the signal obtained by rotating the analyzer was a symmetrical sine wave.

As seen in Figure 7.5, the angle of the extinction position increased as the pH decreased so that, as confirmed independently by Yeh et al. (1987) using a polarized-laser ellipsometer, muscle birefringence increases as pH is decreased (Figure 7.6), but problems also arose from the effect of pH on muscle fiber diameter. There was an approximately 15% decrease in diameter as pH decreased from about pH 7 to 5.5. Because of this bias, the changes in path difference shown in Figures 7.5 and 7.6 are underestimates of the path difference for a constant thickness of muscle.

7.4 Tilting Stage and Polarization of Reflected Light

While the conventional polarizing LM already is appreciated for its many applications in materials science (Piller, 1979), it could be used more widely in many aspects of research, from bread making (Marchant and Blanshard, 1978) to diagnostic pathology (Whittaker et al., 1994). For any sample small enough to be placed upon its stage, the polarizing LM can be used as a compact optical bench, allowing a variety of novel experiments. For example, many solid objects and food commodities are both translucent and glossy, and separating the two is important, as in the quality control of extruded plastics and painted surfaces (Judd and Wyszecki, 1975). Reflected light may contain diffuse, sub-surface scattering, plus glossy surface reflectance following Fresnel equations for polarization of reflections at a boundary (Judd and Wyszecki, 1975). If a commodity is heterogeneous in composition, each component may have its own optical properties, but these are averaged by conventional techniques of colorimetry, which typically integrate an area of several square centimeters. This is an ideal problem for investigation with the polarizing CAM.

There are, however, special problems in the examination of biological tissues with a tilting stage (Swatland, 1995). From Brewster's law, the polarization angle (θ) of reflected rays is related to refractive index (n), n = tan θ, but the refractive index of tissue fluids may be quite variable. Furthermore, whereas overall surface reflectance may indicate sample boundary conditions, such as surface wetness or smoothness, its spatial pattern may indicate intrinsic structure, such as plant cell walls, globular fat droplets, or biological fibers. A final complexity is that tissues such as muscle, as well as larger biological structures such as feathers, nacreous shell layers, and chitinous exoskeletons, may sometimes exhibit natural iridescence from destructive interference.

7.4.1   Hardware

The CAM with polarizing accessories described earlier in this chapter was combined with the tilting stage described in Chapter 6, as shown in Figure 7.7. Above the analyzer was a Zeiss MPM-03 photometer head with a series of measuring apertures (Zeiss 477321). Above this was a depolarizer (Zeiss 453653) to prevent polarized light from interacting with the grating monochromator and photomultiplier. Above the grating monochromator (Figure 7.7, 6; Zeiss 474345) was a filter changer (Figure 7.7, 7; Zeiss 477215) to remove higher-order harmonics from the grating output. The tilting stage (Figure 7.7, M2) was operated at increments of 0.45° with the top of the tilting stage 10 mm below the axis of rotation. Shims were used to elevate the top of the sample to the axis of rotation. The whole assembly (stepper motor plus tilting stage) was mounted on a rotary, gliding stage (Zeiss 473454) to facilitate the centering and rotation of the sample in the field of view. An optical fiber (Figure 7.7, 4) attached to the tilting stage and tapping into the illuminator via a mirror (Figure 7.7, 3) was used to illuminate the specimen.

7.4.2   Theory

In Figure 7.8, the area of the specimen illuminated by the optical fiber (Figure 7.7, 4) is shown as an ellipse, with a long axis, AB. The axis of rotation of the tilting stage corresponded to the short axis (Figure 7.8, CD) of the illuminated ellipse. Thus, the optical axis of the optical fiber (OF) and the long axis (AB) of the illuminated ellipse defined the plane of incidence. The light from the optical fiber was unpolarized, and the central axis of the cone of illumination was at a fixed angle

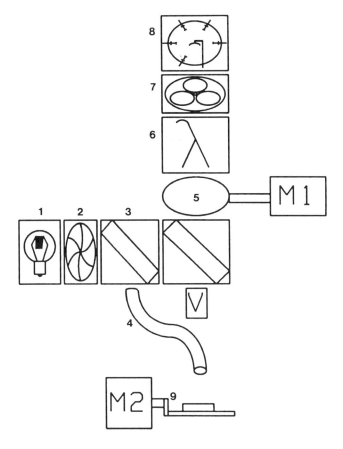

FIGURE 7.7
Block diagram of polarizing CAM with tilting stage, showing illuminator (1), shutter (2), and mirror (3) to deflect light into an optical fiber (4); analyzer (5) and motor (M1); monochromator (6); stray-light filter (7); photomultiplier (8); tilting stage (9) with motor (M2); and vertical illumination through the objective (V).

of 45° to the specimen. In a simple situation, at a flat interface between dielectric media differing in refractive index, one would expect the reflected rays (Figure 7.8, R) to be polarized primarily at 90° to the plane of incidence, while the rays refracted and transmitted into the sample (Figure 7.8, T) should be polarized primarily in the plane of incidence.

In Figure 7.8, the orientation of the microscope objective is shown by the two arrows: at 0° stage tilt and at 45°. Although the optical fiber was fixed with respect to the stage, the whole assembly could be rotated on the glide stage on which it was mounted, so either of the AB or CD axes could be aligned with the 0° axis of the analyzer. The analyzer was at 0° when in the W-E axis of the microscope (because normally it is used for extinction measurements against a substage polarizer set at 0°

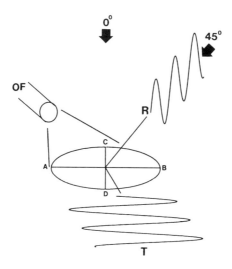

FIGURE 7.8

Polarization of reflected light. The optical fiber (OF) illuminated an elliptical area on the sample (axes AB and CD). Reflected light (R) was polarized at right angles to the plane of incidence, while transmitted light (T) was polarized in the plane of incidence. The arrows show the position of the microscope objective at 0° and 45° tilt of the stage.

in the N-S axis). Except for Figure 7.10, the axis of rotation of the tilting stage was W-E with respect to the microscope.

7.4.3   Illumination and Tilting Restraints

Design restraints were imposed by the limited space for the optical fiber to illuminate the sample laterally. Thus, it was only possible to measure light reflected from 45° to 90° with respect to the angle of incidence. Only at full tilt of the stage (Figure 7.8, 45°) was the angle of incidence equal to the measured angle of reflection. Fortunately, polarization of reflected rays was already detectable by ≤45°. Stronger polarization should have occurred at >45°, but this could not be measured because the microscope could not be focused adequately. Even with a relatively large measuring aperture (1.25 mm), it was possible to keep the high and low points of the measured field within the focal depth of the objective. But, a light microscope cannot be focused adequately on a plane that is approaching the vertical.

An alternative illumination pathway was possible using the built-in vertical epi-illumination system of the microscope. At 0° tilt with a flat, optically simple sample and a vertical epi-illuminator, one would expect the light re-entering the microscope objective to be dominated by surface reflectance (relatively unpolarized with incident light normal to the sample surface), plus some diffuse light from beneath the sample surface. At a 45° tilt, one would expect the surface reflectance to be reduced (reflected away from the optical axis) but diffuse light to remain about the same.

7.4.4   Extinction Coefficient

The efficiency of polarizers and analyzers in a microscope may be checked by measuring their extinction coefficients (k) with transmittance (T) at different wavelengths:

$$k = \log_{10} (T_0 / T_{90})$$

where T_0 is with the analyzer parallel to the polarizer, and T_{90} is with the analyzer perpendicular to the polarizer. A similar approach was used to find the degree to which light reflected from samples was polarized, using reflected (R) light rather than transmitted light and replacing the polarizer by the sample. Thus,

$$k = \log_{10} (R_0 / R_{90})$$

so that a high extinction coefficient indicates a strong polarizer, and vice versa. A subscript is used to denote the angle of tilt at which k was measured.

7.4.5   Intrinsic Anisotropy

When assembling a polarizing CAM, there is always a risk that some unseen factor has created internal optical anisotropy which, if accepted uncritically, might wrongly be attributed to the sample. If a diffuse white sample, such as opal glass, is illuminated from above with nonpolarized light through an epi-condenser, one would not expect rotation of the analyzer to cause much change in reflectance detected by the photometer. If this did occur, the source of the unwanted polarization could be either before the analyzer (so that light reaching the analyzer was partly polarized) or after the analyzer (so that the analyzer was polarizing the light, which then was interacting with an optically anisotropic component in or beneath the photometer). Opal glass with very low gloss and Teflon (polytetrafluoroethylene) with appreciable gloss were used as white samples to check for intrinsic optical anisotropy in the microscope. The optical properties of Teflon as a white standard are described by Weidner and Hsia (1981).

7.4.6   Vertical 0° vs. Lateral 45° Illumination

With vertical illumination (through the built-in epi-condenser of the microscope), the photomultiplier was set to an appropriate gain and amplification with the analyzer at 0°. Then the analyzer was rotated from 0° to 180° in steps of 10°, and the photometer reading at each angle was used as 100% reflectance. A slight degree of anisotropy ($k_0 \approx 0.05$) in the system was detected from the photometer readings, which was canceled by using these photometer readings as 100% reflectance at each angle. Then, when the stage was tilted, it was possible to test for any change in degree of

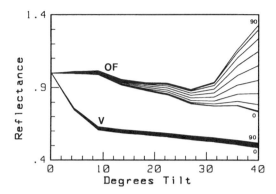

FIGURE 7.9

Effect of tilting a Teflon standard when using vertical (V) illumination through the epi-condenser compared with lateral illumination at 45° through the optical fiber (OF). Results with analyzer angles from 0° to 90° are shown.

polarization relative to a 0° tilt. This was found to be very low. When the photomultiplier was reset with lateral illumination through the optical fiber (instead of through the epi-condenser), anisotropy was reduced ($k_0 \approx 0.006$), thus implicating the vertical illuminator as the source of the intrinsic anisotropy. Consequently, whenever the illumination pathway was changed (from optical fiber to vertical illuminator and vice versa), the system was re-standardized (on a diffuse white standard at a 0° tilt). This was facilitated by storing two matrices on disk, one with settings for vertical illumination and one for lateral illumination.

With vertical illumination (Figure 7.7, V), as the stage was tilted, reflectance decreased because the illumination was reflected away from the optical axis of the microscope. Figure 7.9 shows this effect for a Teflon standard. As the tilt angle increased, polarization became detectable ($k_{40.5} = 0.03$). When the same Teflon standard was illuminated laterally at 45° through the optical fiber, the decrease in reflectance with tilting was greatly reduced, but the strength of polarization showed a strong increase with tilt angle ($k_{40.5} = 0.26$). Opal glass did not show such strong polarization as Teflon when tilted with lateral illumination, because it had less surface reflectance.

For lateral illumination, too much of the dynamic range of the photomultiplier was used when it was standardized at a 0° tilt, so that it became saturated at 45° (overwhelming the normal safety factor used, so that data > 40° are not available for Figure 7.9). Thus, when standardizing the photomultiplier, it became necessary to tilt the stage to find the maximum reflectance in any plane of polarization, then to reduce the proportion of the dynamic range used for standardization at 0° tilt standardization. It seemed unwise to depart from using 0° tilt as the basis for standardization because this would have complicated the comparison of different samples.

With the axis of rotation of the tilting stage (Figure 7.8, CD) aligned in the 0° N-S axis of the microscope, surface reflectance was polarized at 90° to the plane of incidence and, thus, at 90° to the optical axis of the microscope. But the optical axis

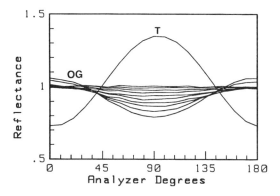

FIGURE 7.10

Effect of rotating the sample 90° about the optical axis of the microscope (all with lateral illumination at 45°). As opal glass (OG) was tilted from 0° to 45° in steps of 4.5°, reflectance increased at 0° and 180° analyzer degrees and decreased at 90°. A Teflon (T) standard rotated 90° about the optical axis of the microscope and tilted to 45° gave the opposite effect.

of the microscope is defined by the substage polarizer (not used in this experiment but normally in the N-S axis), to which the primary axis analyzer is at 90°. Thus, the strongest reflectance was at 0° and 180° analyzer angle. This is shown for an opal glass (OG) standard in Figure 7.10. When the sample was rotated by 90° (using the gliding stage that supporting the tilting stage), the relationship was reversed, as shown by the single line for maximum polarization of a Teflon standard (Figure 7.10, T). Figure 7.10 also demonstrates the stronger polarization effect of Teflon relative to opal glass.

7.4.7   Alignment and Depth of Focus

It was difficult to adjust specimen position and height so that the area of interest was always at the three-way intersection of the optical axis of the microscope, the axis of rotation of the tilting stage, and the axis of the cone of illumination. When the alignment was inadequate, the cone of illumination traveled across the area of interest, or the area of interest moved out of the optical axis of the microscope as the sample was tilted. When specimens were properly adjusted, the area of interest did not move with respect to the photometer aperture when the stage was tilted. The ×4 objective in use had a considerable depth of focus so that there were only minor changes in focus of the high and low edges of the field.

7.4.8   Surface Irregularities

There were undulations at various tilt angles, especially when using lateral illumination of a Teflon standard. Microscopic wrinkles in the Teflon tape appeared as bright

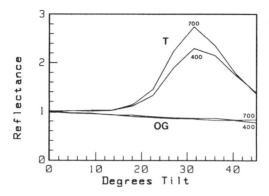

FIGURE 7.11
Effect of tilting opal glass (OG) and Teflon (T) standards on reflectance at 400 and 700 nm (analyzer = 0°).

lines at certain angles (surface reflectance along the crest of a wrinkle). Thus, as the sample was tilted, bright lines appeared transiently in the area of interest if tilt angle directed specular reflectance to the photometer.

7.4.9   Spectrophotometry of Tilted Standards

Standardization of the photometer at each wavelength (from 400 to 700 nm in steps of 10 nm) was done in the same way as for polarimetry (either on opal glass or Teflon, both at 0° tilt and 0° analyzer). In this configuration, diffuse light made a major contribution to the light reaching the photometer. When opal glass (low gloss) was tilted, overall reflectance (diffuse + surface) at 400 nm and at 700 nm decreased to the same extent (Figure 7.11, OG). However, if part of a Teflon standard with strong surface reflectance filled the measuring aperture, reflectance increased as the stage was tilted, reaching a maximum when the angle of reflection was similar to the angle of incidence (bearing in mind that, at the microscopic level, the Teflon standard was not perfectly flat so that the macroscopic tilt angle was only an approximate guide to the angle of the microscopic sample surface). At maximum reflectance (diffuse + surface), reflectance was higher for light at 700 nm than at 400 nm. Thus, at standardization, a relatively high intensity of light at 400 nm had reached the photometer (because scattering tends to be inversely proportional to the fourth power of wavelength). But, when surface reflectance was added as the stage was tilted (all visible wavelengths reflected equally), the ratio of diffuse to reflected light was changed. Thus, the proportional increase (surface/diffuse) was higher for light at 700 nm than for light at 400 nm.

7.4.10   Spectrophotometry of Tissue Surface Reflectance

In a turbid medium containing a myriad of scattering particles, such as a translucent biological tissue, the light path through which absorbance occurs is greatly increased by scattering (Butler, 1962). Thus, the degree of absorbance becomes an interaction between the concentration of the chromophore and the lengthening of the light path by scattering. In addition, for tissues such as skeletal muscle, surface reflectance may be anisotropic, with a higher reflectance when the incident illumination is perpendicular to the muscle fibers than when it is coaxial. Unfortunately, the pioneer work in reflectance spectroscopy of intact tissues by Ray and Paff (1930) did not receive the attention it deserved, and it was not until the advent of fiberoptic spectroscopy that work was resumed in this area. At present, the possibilities of surface reflectance spectroscopy with the CAM are almost completely ignored by biologists.

Myoglobin is the dominant chromophore of skeletal muscle and determines the reflectance spectrum of muscle (Theorell and de Duve, 1947; Naughton et al., 1957). When myoglobin and its derivatives have been removed by washing, reflectance is almost a linear function of wavelength (from about 0.3 at 420 nm to about 0.75 at 700 nm). The Soret absorbance bands for deoxymyoglobin (purple-red), oxymyoglobin (bright-red), and oxidized metmyoglobin (brown) are at 434, 416, and 410 nm, respectively (Bowen, 1949). Deoxymyoglobin has an absorbance band at 555 nm that is replaced in oxymyoglobin by a strong absorbance band at 578 nm and a slightly weaker band at 542 nm, although metmyoglobin formation generally masks this difference between reflectance at 542 and 578 nm. The relatively high myoglobin concentration of dark red skeletal muscle reduces the overall intensity of reflectance spectra to about one third that of washed muscle or a white muscle lacking myoglobin.

The microscope was standardized on a sample of dark red muscle (to give reflectance $\approx$ 1 from 400 to 700 nm) at 0° tilt and 0° analyzer (Figure 7.12, 1). When the stage was tilted to 45°, reflectance increased at 420 nm and at 550 and 580 nm (Figure 7.12, 2). When the analyzer was rotated to 90° (thus removing most of the surface reflectance added when the stage was tilted), reflectance returned almost to initial values (Figure 7.12, 3).

Line 2 of Figure 7.12 has almost the same shape as an absorbance spectrum of oxymyoglobin. This may be explained by an effect similar to that seen in Figure 7.11. In diffuse light, some wavelengths are less intense than others because scattering is a function of wavelength (as in Figure 7.11), or because of selective absorbance of certain wavelengths (as in Figure 7.12). When surface reflectance is added to diffuse light by tilting a sample, its spectral composition is determined by the emission spectrum of the illuminator, thus adding light at wavelengths that may be weak in the diffuse light used for standardization. Hence, line 2 in Figure 7.12 followed the absorbance spectrum of oxymyoglobin, the dominant myoglobin derivative in the muscle. Removing much of the surface reflectance by rotating the analyzer caused reflectance to return near to the values at standardization (Figure 7.12, 3).

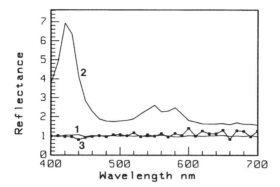

FIGURE 7.12
Effect on spectrophotometry of tilting skeletal muscle: sample standardized at 0° tilt and 0° analyzer, line
1; sample tilted to 45°, line 2; and analyzer rotated to 90°, line 3. The longitudinal axes of the muscle fibers
were parallel with the axis of rotation of the tilting stage.

7.4.11   Spectrophotometry of
Diffuse Reflectance

Following from Figure 7.11, one would expect light at 700 nm to be slightly more
intense than light at 400 nm for line 3 of Figure 7.12. This was tested with least-
squares regressions. Whereas the standardization line (Figure 7.12, 1) was not
significantly different from the horizontal, line 3 of Figure 7.12 had a positive slope
(8.2 E-4 reflectance per nm, $P < 0.005$). Thus, although this wavelength-related effect
was very small relative to that caused by selective absorbance, it was definitely
detectable.

7.4.12   Polarimetry of Surface Reflectance

Muscle, tendon, cartilage, and adipose tissue were measured, searching for subtle
effects related to refractive index and birefringence; however, it soon became apparent that the dominant factor in any measurement was the three-dimensional structure
of the sample surface. Hyaline cartilage was white and translucent, and had a low
gloss. For samples with a flat cut surface, the polarimetry data obtained on tilting
were very similar to those of the Teflon standard (for cartilage, $k_{45} \approx 0.17$). For cut
surfaces of adipose tissue, rounded cells were visible, but overall surface reflectance
was low, $k_{45} \approx 0.07$.

In contrast to these predictable results from tissues such as cartilage and adipose
tissue (which could be cut to have a flat surface), muscle and tendon produced less
predictable results. The fibers of both muscle and tendon are bound into a fascicular
structure, creating a directional grain on the sample surface. When an area for
measurement was relatively flat, the results of tilting the sample were predictable

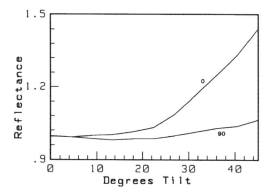

FIGURE 7.13

Effect on polarimetry of tilting a smoothly cut specimen of skeletal muscle: reflectance at 0° and 90° analyzer, with the longitudinal axes of the muscle fibers perpendicular to the plane of the stage.

(Figure 7.13); however, when bundles of either muscle fibers or collagen fibers rolled through the optical axis, they gave a flash of surface reflectance into the optical axis and a corresponding flash of strongly polarized surface reflectance. The effect was strongest when the fibers were parallel with the axis of rotation of the tilting stage (Figure 7.14). Apart from this effect, polarization was generally low in areas of the muscle and tendon with a low surface reflectance and high in areas with high surface reflectance ($k_{45} \approx 0.14$ and $k_{45} \approx 0.68$, for dull and glossy areas, respectively).

Muscle and tendon specimens with both an appreciable gloss and a fascicular grain acted as partial polarizers, even at 0° tilt ($k_0 \approx 0.05$). In other words, the sides of fibers protruding from the sample surface already were tilted relative to the optical axis of the microscope, so that reflected light was partly polarized.

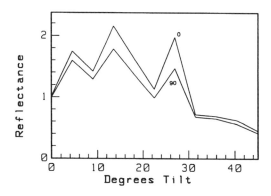

FIGURE 7.14

Effect on polarimetry of tilting a specimen of tendon with a surface grain caused by internal fascicular structure: reflectance at 0° and 90° analyzer, with the longitudinal axes of the bundles of collagen fibers parallel to the axis of rotation of the tilting stage.

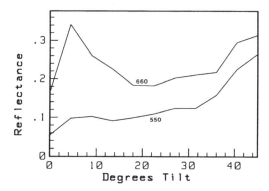

FIGURE 7.15

Effect of tilting on second-order interference green (550 nm) and orange (660 nm) iridescence in a single muscle fiber, cut perpendicularly to the longitudinal axes of the muscle fibers.

7.4.13   Iridescence

In iridescent samples of skeletal muscle with no unusual features with respect to gloss ($k_{45} \approx 0.4$), the flashing of interference colors was detectable on tilting, as in the sample shown in Figure 7.15, where there was a flash of second-order orange (660 nm) but not of second-order green (550 nm) at a 4.5° tilt. The tilting of iridescent feathers produces similar but more complex results.

7.5   Why Is the Polarized Light Microscope Seldom Used by Biologists?

Despite the efforts of Bennett (1950), Oster (1955), and Ruch (1966), which are still worth reading today, polarized LM seems to have been lost in the rush towards electron microscopy. POL objectives and accessories are only slightly more expensive than normal, they are easily retrofitted to most research LMs, and they are not difficult to use. For almost any birefringent tissue component, its optical properties rank equal in scientific importance and reliability to any other property, as shown in Table 7.2, using collagen as an example.

The birefringence of collagen fibers is not a trivial or random property, because it is determined by molecular structure (Romhányi, 1986). In addition, the polarized LM has a variety of practical applications in identifying pathological abnormalities in collagen structure (Whittaker et al., 1988; Hirshberg et al., 1996), studying wound healing (Wolman and Gillman, 1972; Rabau et al., 1995), and in biomechanics (Ortmann, 1975; Arokoski et al., 1996). It would be inappropriate here to provide an in-depth review of the great variety of birefringent tissue components, but the conclusion would be the same for all, as it is for collagen. Birefringence has both a

TABLE 7.2
Birefringence as a Fundamental Feature
of Collagen Fibers

Histological Name	Collagen Fiber	Reticular Fiber
Location	Tendon, epimysium, aponeuroses, fasciae, etc.	Endomysium around muscle fibers and adipose cells, etc.
Biochemical name	Type I	Type III
Tropocollagen structure	$[\alpha 1(1)]_2 \alpha 2(I)$	$[\alpha 1(III)]_3$
Associated proteins	Proteoglycan	Proteoglycan, fibronectin
Silver staining	Black	Yellow
Fluorescence	Strong	Weak
Fluorescence peak emission	430 to 440 nm	≈ 500 nm
Intrinsic birefringence	Strong	Weak
Sign of intrinsic birefringence	Positive	Negative
Form birefringence	Positive	Positive

basic and applied importance. Thus, why the use of the polarized LM is largely restricted to the study of minerals, synthetic fibers, and opto-electronic components is a mystery. However, perhaps changes are on the way. The Pol-Scope developed at the Marine Biological Laboratory in Woods Hole, MA, uses circularly polarized light so that birefringent structures appear equally bright at all orientations (Oldenbourg, 1996). Electro-optical modulators are used as compensators, so that path differences may be found electrically rather than electromechanically, as described in this chapter. Thus, ellipsometry measurements can be made on any structure in a whole-field video, rather than on part of the field defined by an aperture.

Chapter 8

Fluorescence

8.1   Introduction

Fluorescent materials absorb radiant energy and then, almost instantly, re-emit some of the energy, usually at a longer wavelength. Primary fluorescence (autofluorescence) occurs in many minerals, such as fluorspar, hence the origin of the word fluorescence. Flavoproteins (Kunz and Gellerich, 1993) and plant cell wall materials, such as lignin (Ames et al., 1992), exhibit strong primary fluorescence, as do fibers of cotton and paper (depending on how they have been processed), as well as wool (Davidson, 1996). Thus, structures such as articular cartilage, heart valve, and lung also are fluorescent because of the connective tissue fibers they contain (Swatland, 1988). Far weaker, however, is the autofluorescence of the flagellum, although even this can be measured by microspectrophotometry (Kawai et al., 1996).

Laboratory dust derived from clothing and human skin is strongly fluorescent. Thus, a CAM always should be kept covered when not in use. Dust on lenses can be a major source of background noise for a fluorescence CAM, which is why the whole optical system must be kept very clean for quantitative fluorescence microscopy.

Primary fluorescence may be quenched (extinguished) by histological reagents containing mercury, iron, and iodine. Thus, after routine processing for light microscopy, initial autofluorescence may be lost, so that the characteristic fluorescence of a stained structure originates secondarily from the stain, such as eosin on elastic fibers (Decarvalho and Taboga, 1996). In biological tissues in their native state, quenching may occur with prolonged or intense excitation. Quenching is reduced at low temperatures (Tiffe and Hundeshagen, 1982).

Secondary fluorescence occurs when a material binds one of the numerous fluorescent dyes or fluorochromes listed by Rost (1995), such as acridine orange or auramine. Or, secondary fluorescence may originate from a fluorescent-labeled antibody (immunofluorescence) or a fluorescent antitumor antibiotic (Crissman et al., 1976). Visible staining for routine histology requires binding of substantial quantities of dye. This may be facilitated by a mordant, such as alum, precipitating the dye onto the target. However, with fluorescence microscopy, very low concentrations of bound fluorochromes are detectable, so that fluorochrome binding generally is more sensitive than dye staining. Immunofluorescence adds a further substantial gain in sensitivity, as well as in specificity. Just as de-staining is an important step in obtaining maximum discrimination with many ordinary histological methods, de-staining of fluorescent dyes may be advantageous (Heldal et al., 1996). In some techniques, time-based measurements of quenching may be used to acquire information about the identity or molecular structure of the source of the fluorescence (Tian and Rodgers, 1991; Peters and Scholz, 1991; Morgan and Mitchell, 1996).

Other applications of fluorescence microscopy include measurement of intracellular pH from the emission spectrum of an indicator dye, such as fluorescein (Visser et al., 1979), detection of intracellular calcium ion movements (Vergara et al., 1991), and the use of voltage-sensitive fluorescent dyes for the detection of membrane depolarization (Loew, 1991; Beach et al., 1996). In neurophysiology, fluorescent dyes also may be used to mark individual cells from which electrophysiological measurements have been made with a micropipet (Chowdhury, 1969; Berthold et al., 1979; Hartveit, 1996). Green fluorescent proteins isolated from marine organisms now are being used as a convenient way of monitoring gene expression (Yang et al., 1996).

Programming for fluorescence microscopy may require a combination of imaging with spectroscopy so that a variety of different color fluorescent markers can be identified in an image (Waggoner et al., 1996). Fluorescence emission spectra of important tissue types, such as the different biochemical types of collagen, may vary considerably in shape when measured at different levels of magnification: at the microscopic level, at the semi-micro level using fiberoptics, or at the macroscopic level using a large-aperture fluorometer (as in quality control specifications for paper manufacturing). Unfortunately, fluorescence measurements are difficult to standardize, and they tend to reflect the apparatus with which they are recorded as much as the nature of the sample (Bashford, 1987). It is also important to be aware that the main peaks, and sometimes the shapes, of excitation and emission spectra may be modified by organic solvents.

The principles and methodology of LM photometry for a manually operated system are well known (Piller, 1977); however, computer-assistance introduces some new possibilities since it enables fluorescence spectra to be measured quite rapidly, compared to manual operation. An accepted manual method for the measurement of relative spectral fluorescence intensities (Zeiss, 1980) may be used as the main algorithm. This chapter describes further technical factors involved in the interaction of optics, sample, and software, using collagen as an example.

8.2   Excitation vs. Emission

Light causing fluorescence is called fluorescence excitation, while light re-emitted is called fluorescence emission. Thus, there are two types of fluorescence spectra — excitation and emission. The main peak of a fluorescence excitation spectrum shows

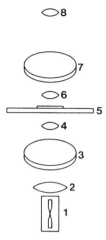

FIGURE 8.1

Main components for LM dia-excitation: arc lamp, 1; collector lens, 2; exciter filter, 3; substage condenser, 4; sample, 5; objective, 6; barrier filter, 7; and ocular, 8.

the wavelengths causing the strongest fluorescence, while the main peak of a fluorescence emission spectrum shows the dominant wavelengths at which light is re-emitted. When creating y-axis labels for graphics and data storage it is important, therefore, to code excitation and emission spectra differently. For example, FLU EXC 450 might be used to indicate a fluorescence excitation spectrum measured at an emission wavelength of 450 nm, using FLU EMS 365 to indicate a fluorescence emission spectrum with peak excitation at 365 nm. Short titles are preferable to long ones to facilitate searching and to leave space for a temporary scaling indicator on a graphics label, such as ×100.

8.3   Dia-Excitation

Old fluorescence LMs usually operate on the principle of dia-excitation, with the excitation passing through (dia-) the microscope slide, as in the normal LM. The main components added to the basic LM are usually an arc lamp (Figure 8.1, 1), an exciter filter (Figure 8.1, 3), and a barrier filter (Figure 8.1, 7). Ideally, the exciter filter should be a low-pass filter giving high transmittance of the main peak of the excitation spectrum with a sharp cutoff for higher wavelengths, while the barrier filter should be a long-pass filter with a sharp cutoff for all the excitation spectrum. This will result in a completely black background against which to display weak fluorescence; however, most filters in common use are less than ideal, as shown in Figure 8.2. Thus, filter spectra are very important for quantitative studies and, if they cannot be found from documentation, they must be measured empirically. An important point to check is the extent of the crossover between exciter and barrier filters.

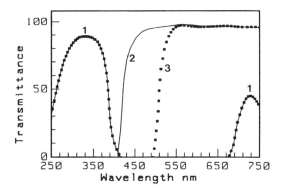

FIGURE 8.2
Transmittance spectra of filters used for dia-excitation: exciter filter UG5 (1) and long-pass barrier filters at 410 nm (2) and 500 nm (3).

Thus, in Figure 8.2, the exciter filter and both long-pass filters transmit red light towards 700 nm. Even more critical, because it may appear as pseudofluorescence, is the small crossover where the longest wavelengths transmitted by the exciter filter may pass through as the shortest wavelengths transmitted by the barrier filter.

8.3.1   Ultraviolet Safety

Most early fluorochromes required UV excitation, but, as immunofluorescence developed into a major method for medical diagnosis, fluorochromes were developed with their excitation maxima at longer wavelengths (typically blue), which enabled a quartz halogen illuminator to be used instead of an arc lamp. With UV arc lamps, there was often a compromise between optical efficiency and safety.

Ordinary glass lenses have only a low transmittance of UV light. Thus, quartz collectors and quartz substage condensers sometimes were used to maximize the UV excitation reaching the sample, especially for expensive LMs also capable of UV absorbance measurements. This increases the risk of UV light reaching the eyes of an operator who unwisely removes the barrier filter from the light path. This is a very important safety point for anyone improvising with an old dia-excitation LM. If in doubt, play safe, and fit the oculars with a UV barrier.

8.4   Epi-Excitation

A dichroic mirror has a metallic coating that reflects some wavelengths but transmits others. In the Ploempak epi-illuminator, developed by J. S. Ploem and first manufactured by Leitz, a dichroic beam splitter is used to produce a dramatic improvement in the layout of the fluorescence LM. Instead of the excitation being transmitted

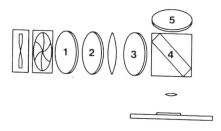

FIGURE 8.3

Main components for LM epi-excitation: heat filter, 1; red attenuation filter, 2; exciter filter, 3; dichroic mirror, 4; and barrier filter, 5.

through the specimen, it is directed downwards through the objective lens onto the top of the specimen (epi-excitation). Thus, the objective acts as both condenser and objective, allowing maximum use of numerical aperture and producing intense excitation focused exactly on the field of view. The Ploempak concept was applied above the level of the objective nosepiece, but a dichroic mirror together with exciter and barrier filters may be miniaturized and located within the objective, using a fiberoptic light guide connected to the objective as the illumination pathway, as in the Makler fluorescence objective. Thus, an ordinary LM may be converted to a fluorescence LM easily and inexpensively but at a fixed magnification.

The main components required for epi-excitation are a heat filter (Figure 8.3, 1), because the metallic film of the dichroic is quite vulnerable, and a red attenuation filter (Figure 8.3, 2) to remove red light (see Figure 8.2, 1). The excitation filter (Figure 8.3, 3) is similar to that used in dia-excitation (Figure 8.2, 1), although now it may be thinner with higher transmittance. A dichroic mirror (Figure 8.3, 4) is selected to reflect the excitation downwards through the objective and onto the specimen. Light transmitted through the dichroic from the illuminator may be focused onto a screen built into the microscope as a convenient way of centering the illuminator. Fluorescence emission from the specimen is captured by the objective and transmitted straight through the dichroic mirror, through a barrier filter (Figure 8.3, 5), and onwards and upwards into the optical axis of the CAM.

Thus, epi-excitation works perfectly well on thick sections or dissected fibers since it is primarily a surface measurement. Quenching proceeds from the surface of the specimen into its depth, so that the fluorescence produced by epi-excitation may be more stable than that detected in an equivalent thin section using dia-excitation. Thus, large-diameter fibers may continue to fluoresce for longer than small-diameter fibers, because the core of the larger fiber takes longer to quench. Rate of quenching also may be indicative of changes in molecular structure (Davydov et al., 1996).

For the CAM, where usually the primary purpose is to measure excitation and emission spectra, the exciter and barrier filters may be replaced by excitation and emission monochromators (Figure 8.4, 3 and 4). A solenoid-operated shutter (Figure 8.4, 1) is required in addition to the illuminator aperture. The stray-light filters on the monochromators should be selected with the same logic used for exciter and barrier filters, that is, short-pass for excitation and long-pass for emission.

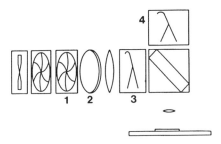

FIGURE 8.4

Main components for CAM epi-excitation: illuminator shutter, 1; heat filter, 2; excitation monochromator, 3; and emission monochromator, 4.

8.5   Importance of Collagen Autofluorescence

Most texts on fluorescence microscopy are concerned primarily with secondary fluorescence, with autofluorescence being relegated to the status of a background nuisance. To balance the books, the examples used here relate to autofluorescence. The general conclusions are equally applicable to secondary fluorescence, but they enable us to side-step the complexities of fluorochrome chemistry and immunofluorescence. Also, for the reasons given below, autofluorescence is beginning to assume an importance of its own, at least for connective tissue fibers and plant cell walls.

The tropocollagen molecule has three alpha chains, but 19 unique alpha chains have been identified, giving rise to 11 biochemical types of collagen. These may be categorized into three general classes: (1) molecules with a long (≈300 nm) uninterrupted helical domain, (2) molecules with a long (≥300 nm) interrupted helical domain, and (3) short molecules with either a continuous or an interrupted helical domain (Miller, 1985). The following types of collagen are used as examples in this chapter. Type I collagen fibers are striated when stained with heavy metals for examination by transmission electron microscopy. They are between 80 and 160 nm in diameter and occur in blood vessel walls, tendon, skin, and muscle. Type III reticular fibers are found in muscle endomysium, around adipose cells, and in the spleen.

Far-UV fluorometry of tyrosine, proline, and hydroxyproline in homogeneous collagen solutions is used for biochemical research on the structure of collagen (Bellon et al., 1985; Na, 1988). Histochemically, fluorescence microscopy may be used to separate elastin and collagen fibers after treatment with phosphomolybdic acid (Puchtler et al., 1973). Autofluorescence also may be used to separate some of the major biochemical types of collagen. Biochemical types I and III collagen have different fluorescence emission spectra (Swatland, 1987). Maximum excitation is near 370 nm and, for practical purposes, may be obtained with the 365-nm peak of a mercury lamp. Type I fibers emit a pre-quenching spectrum for longer than type III fibers, probably because type I fibers have a larger diameter than type III fibers.

The three polypeptide strands of tropocollagen are linked by stable intramolecular bonds originating in the non-helical ends of the molecule, and stable disulfide bonds between cystine molecules occur in the triple helix. Much of the high tensile strength of collagen fibers originates from covalent bonds between adjacent tropocollagen molecules. During animal growth, covalent crosslinking increases, and collagen fibers become progressively stronger and less elastic. Many of the intermolecular crosslinks in young animals are reducible while, in older animals, reducible crosslinks are converted to nonreducible crosslinks. Pyridinoline is a nonreducible fluorescent crosslink of collagen, and its presence in the urine is used as a marker of pathological collagen degradation (Steinhart et al., 1994). The fluorescence excitation maximum of pyridinoline is difficult to reach with a fluorescence LM (excitation maximum at 295 nm, with emission from 400 to 410 nm), but the upper portion of the excitation is attainable at 365 nm. With excitation at 335 nm and 370 nm and measurement at 385 nm and 440 nm, respectively, the fluorescence of collagen increases exponentially with age and is a reliable marker for physiological age (Odetti et al., 1994). Similar conclusions have been reached for collagen in lung (Bellmunt et al., 1995) and bone (Tomasek et al., 1994).

Thus, the autofluorescence of collagen is not just a background nuisance, but rather has become a useful indicator of biochemical composition, connective tissue pathology, and physiological age. Neither is the autofluorescence of collagen exceptional in this regard, because similar conclusions may be reached for the autofluorescence of elastin (Suyama and Nakamura, 1992) and autofluorescence of cellular pigments such as lipofuscin (Clokey and Jacobson, 1986).

8.6 Fluorometry Software

8.6.1 Relative Spectral Fluorescence Intensity

Different optical layouts for this chapter were all standardized in the same way for the determination of relative spectral fluorescence intensities (Zeiss, 1980). From a software perspective, standardization of a CAM for fluorometry is far more complex than for spectrophotometry. In essence, spectrophotometry with a CAM is very simple because it is based on the ratio of light in to light out. Thus, if the standardization protocol is reasonable, moderate differences in the emission spectrum of the illuminator are unimportant. Fluorometry, on the other hand, has many of the problems associated with true radiometry, measuring the emission spectrum of an unknown source. The logic rapidly becomes circular. Radiometry can be done if the relative spectral sensitivity of the photometer is known, but how can this be determined without a source with a known emission spectrum? Fortunately for the average scientist, this conundrum has been solved by meticulous physicists who, via our various national standardization organizations, provide us with calibrated sources and photometers.

Fortunately for the CAM programmer, most of the emission spectra required by CAM users are in the visible spectrum, which enables us to use a halogen source with a known emission spectrum (Chapter 2) to "show" the PMT what white light looks like. Thus, the spectrophotometer scans from 400 to 700 nm, usually in steps of 10 nm. The photometer reading at each wavelength is weighted by a factor corresponding to the emission spectrum of the source. For example, comparing only two wavelengths for the sake of simplicity, if emission at the first is only one tenth that at the second, the photometer response for the first wavelength is multiplied by 10 (while the photometer response for the second wavelength is multiplied by one). Thus, if this system were to be exposed to a hypothetical source with equal intensity at all wavelengths, the PMT would read the same for both wavelengths. Software may be tested by measuring the fluorescence emission spectrum of uranyl glass and comparing it to the manufacturer's calibration (given for uranyl glass in Zeiss, 1980). Other standards are optical fibers doped with uranium or europium (Velapoldi et al., 1974) or fluorescent pigments in a mounting medium (Walker and Watts, 1970).

This protocol cuts quite a few corners and does not allow fluorescence spectra to be expressed in radiometric units or as quantum yields. Thus, the spectra are only relative spectra and are often normalized (scaled to a maximum value of 1). An alternative approach to the problem is to split the excitation beam, taking a reference beam of about 15% to a quartz cuvette containing Rhodamine B in ethylene glycol. This emits red light in proportion to the intensity of the monochromatic excitation so that fluorescence may be expressed ratiometrically (Mayer and Novacek, 1974; Wreford and Schofield, 1975).

8.6.2   Spectrofluorometry

Despite the great improvements obtained by using epi-excitation rather than dia-excitation, fluorescence often is still very weak (relative to normal bright field microscopy). Dispersing light through a monochromator to measure spectra introduces a further attenuation, often pushing the PMT to the limit of its performance. Consider a typical algorithm for standardizing a spectrophotometer, as in Chapter 5.

[1]   Create an appropriate blank and find the wavelength giving the maximum PMT response (Pmt).

[2]   Adjust the PMT to give a maximum response.

[3]   Temporarily close the illuminator shutter to find dark-field measurement (Nl, no light).

[4]   Go to each wavelength and measure Pmt.

[5]   At each wavelength, subtract Nl from Pmt to get a blank measurement.

[6]   Place a sample in position and go to each wavelength to measure Pmt.

[7]   At each wavelength, subtract Nl from Pmt to get the sample measurement.

[8]   At each wavelength, express the sample measurement as a function of the blank
        measurement.

With an unknown system or one that has been altered, it is necessary to scan through
all the wavelengths to find the right wavelength for step [1] so that the whole
algorithm requires two complete scans with the monochromator before the sample
can be measured. When a long integration time is required for a weak signal, this can
take an appreciable length of time. Also, at the extremes of the spectral range
attempted, a low intensity of light may fall on a relatively insensitive photometer,
especially with fluorometry. Thus, only a small fraction of the dynamic range of the
photometer may be used at these wavelengths. For example, there might be a
thousand gray levels near the center of the spectral range and five at the edges. With
a CAM, however, it is possible to adjust the sensitivity of the photometer so that the
dynamic range can be reset at each wavelength, as follows.

[9]   Create an appropriate blank, go to each wavelength, and adjust the PMT to give a
        standard response (such as 90% of maximum range).
[10]   At each wavelength, use the illuminator shutter to find Nl.
[11]   At each wavelength, subtract Nl from Pmt [9] to get the blank measurement.
[12]   At each wavelength, store the gain, high voltage, blank measurement, and Nl in a
        matrix.
[13]   Place a sample in position.
[14]   Go to each wavelength, reset the PMT, and measure Pmt.
[15]   At each wavelength, subtract L from Pmt to get the sample measurement.
[16]   At each wavelength, express the sample measurement as a function of the blank
        measurement.

With this algorithm, only one complete scan with the monochromator is required
before a sample can be measured. A potential problem, however, is that the photometer response may vary with wavelength so that the linearity of the system may be
in doubt. The first method, [1] to [8], will be called the classical method while the
second method, [9] to [16], will be called the maximum dynamic range method. The
method for the measurement of relative spectral fluorescence intensities requires that
the photometer should first be standardized against a source with a known emission
spectrum. This raises the question of which of the two methods of photometer
standardization is most appropriate (Swatland, 1990b). Is it possible to save time by
using the maximum dynamic range method rather than the classical method?
        The system was standardized using an algorithm that combined both classical
and maximum dynamic range methods. Measurements were made retrospectively in
a way that simulated either method so that the results were based on a single set of
standardization conditions. Step [1] of the classical algorithm was undertaken retrospectively by searching the matrix collected in steps [1] to [4] of the maximum
dynamic range method. The high voltage and gain at the wavelength with the
maximum response were used for all wavelengths.

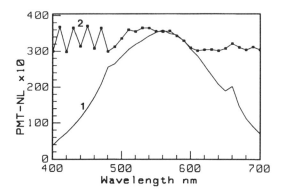

FIGURE 8.5
Blank measurement (raw data for Pmt – Nl) for the classical method (1) and maximum dynamic range method (2).

8.6.3   Comparison of Methods at High Light Intensity

The goal here is to compare classical and maximum dynamic range methods from 400 to 700 nm under working conditions with a high light intensity (100-W halogen source, ×10 objective, 10-nm monochromator bandpass, and 0.16-mm measuring aperture) and rapid standardization and measurement. In biological applications, sample variance may be relatively high and may require replication of samples. Thus, a mean value of a large number of spectra collected rapidly may be of more value than the mean of a small number of samples collected slowly with high precision. The rapid measurement of spectra is required for fluorometry because of fading or fluorescence quenching.

The relative proportions of the dynamic range used in the two methods are shown in Figure 8.5. The blank measurement is shown as Pmt – Nl, as at step [5] of the classical method and step [11] of the maximum dynamic range method. A Pmt of 4000 for this system was a safe maximum for the dynamic range, above which there was a risk of photomultiplier saturation. The magnitude of the safety margin depends on the range of light intensities across the spectrum. For the classical method (Figure 8.5, 1), the maximum light intensity was at 560 nm and 10% of the dynamic range was used at the wavelength with the lowest light intensity (400 nm). For the maximum dynamic range method (Figure 8.5, 2), a relatively wide (20%) acceptance window was used. If the acceptance window is too narrow, there may be wavelengths at which the system cannot be standardized.

The results obtained by placing neutral density filters (3, 12, and 50% transmittance) in front of the light source are shown in Figure 8.6, where lines show the classical method and boxes show the maximum dynamic range method. Neutral density filters produce a uniform attenuation across the visible spectrum by partial reflection. Viewed subjectively, both methods gave a similar result. The results were

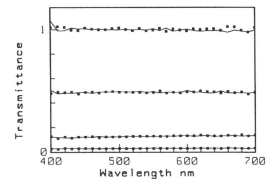

FIGURE 8.6

Classical vs. maximum dynamic range methods of PMT standardization at high light intensity, showing re-measurement of the blank and measurement of 3, 12, and 50% transmittance neutral density filters by the classical method (solid lines) and maximum dynamic range method (boxes).

tested by calculating a least-squares linear regression of transmittance against wavelength (slope = transmittance per nm). All of the biases that were detected were weak and unimportant; however, for remeasurement of the blank, the classical method had a stronger bias (slope –9.46 E-5, P < 0.005) than the dynamic range method (slope = –2.28 E-5, NS). At the lowest level of transmittance, both methods had a very slight bias (slopes of 1.05 E-5 and 1.62 E-5, for classical and maximum dynamic range methods, both P < 0.005). Thus, with a high light intensity, both the classical and maximum dynamic range methods may give similar results.

8.6.4   Comparison of Methods
         at Low Light Intensity

The goal here is to compare classical and maximum dynamic range methods at a low light intensity, as might occur with fluorometry of autofluorescence (×100 objective, 5-nm monochromator bandpass, 0.08-mm measuring aperture). For the classical method, the fractions of the dynamic range in use were similar to those shown in Figure 8.5. For the maximum dynamic range method, the fractions of the dynamic range in use at low wavelengths were reduced (because dark-field measurements were higher).

The results of remeasuring the neutral density filters are shown in Figure 8.7. Viewed subjectively, both methods had a positive bias at low wavelengths. Since the neutral density filters were the same as those used previously (relatively flat transmittance spectra in Figure 8.6), they were not the cause of the bias. However, the classical method was less affected than the maximum dynamic range method and, for remeasurement of the blank, the bias of the classical method (slope = –9.60 E-5, P < 0.05) was less than for the maximum dynamic range method (–4.70 E-4, P < 0.005).

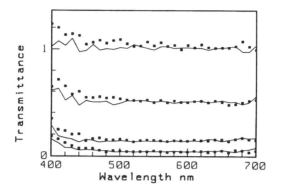

FIGURE 8.7
Classical vs. maximum dynamic range methods of PMT standardization at low light intensity, showing re-measurement of the blank and measurement of 3, 12, and 50% transmittance neutral density filters by the classical method (solid lines) and maximum dynamic range method (boxes).

At the lowest levels of transmittance, the bias of the classical method (-1.22 E-4, $P < 0.005$) was half that of the maximum dynamic range method (-2.48 E-4, $P < 0.005$). Thus, the classical method of standardizing a photometer is preferable for fluorometry where light levels are low.

8.6.5   High vs. Low Light Intensity for Standardization

In the protocol for the determination of relative spectral fluorescence intensities, a comparison source with a known emission spectrum is used to correct for the spectral sensitivity of the photometer. For the CAM, this is normally done with transmitted light from a built-in halogen source, taking care that nothing in the light path (such as a dichroic mirror near its cutoff point) causes any significant changes in the emission spectrum. Thus, the operating conditions of the microscope may differ between photometer standardization and sample measurement.

In the protocol for a manual system (Zeiss, 1980), sample fluorescence is measured first using the full dynamic range of the photometer at the wavelength of maximum fluorescence. Next, the comparison source is measured, also using the full dynamic range of the photometer at the peak wavelength. With a CAM, however, it is more convenient to reverse this procedure so that the system is standardized first, then checked against a standard such as uranyl glass, and only then used to measure samples. The programmer is faced with the problem of deciding what light intensity to use for the comparison source. Should it be at a high intensity or at a relatively low intensity, similar to that expected from the fluorescent sample?

This question may be answered as follows. The fluorescence of teased collagen fibers was measured after standardizing the PMT at three light intensities. At full

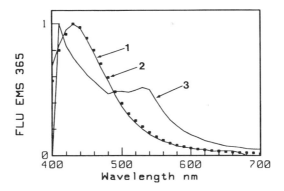

FIGURE 8.8

Fluorescence emission spectra of teased type I collagen fibers using different light intensities from a halogen source to standardize the CAM for relative spectral fluorescence intensity measurements: full intensity, 1; 0.5 transmittance, 2; and 0.38 transmittance, 3.

intensity the light path from a 100-W halogen source was unreduced. Two lower intensities were created with 0.5 and with 0.5 + 0.12 neutral density filters in the light path. The results are shown in Figure 8.8. There was little difference in the fluorescence emission spectrum between standardization at full or 0.5 intensity, but at 0.38 there was an anomalous peak at 410 nm. A similar distortion caused by low light intensities at standardization was also found for an alternative optical system and sample, using uranyl glass connected to the CAM by optical fibers. Thus, it was concluded that the distortion did not originate from the optical separator (microscope dichroic mirror or bifurcated light guide) or from the samples.

Thus, in this case, a relatively high light intensity was used to standardize the photometer even though this may have resulted in a further reduction in the fraction of the dynamic range of the PMT when measuring the fluorescence of a sample.

8.7 Fluorescence Blank and Pseudofluorescence

Figure 8.9 shows four problem spectra of connective tissue fibers. The fibers were teased apart and mounted in distilled water under a cover slip. A small measuring aperture (0.18-mm diameter with ×100, NA 1.3 neofluar) was used to ensure that the measured field was evenly filled. An epi-fluorescence condenser (Zeiss IIIRS) was used with a UG1 (Zeiss 467968) excitation filter as the excitation monochromator and a dichroic mirror (FT395) as the separator. Respectively, spectra 1 and 2 are for chemically purified fibers of types I and III collagen (Sigma Chemical Company; St. Louis, MO) that dominate the tissues shown by spectra 3 and 4, which are for extramuscular tendon and intramuscular connective tissue. Although the tissue types are not pure (extramuscular tendon may contain type III collagen and intramuscular connective tissue may contain type I collagen), there is an appreciable

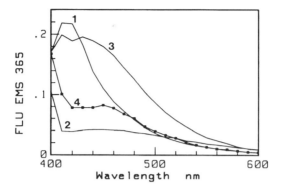

FIGURE 8.9

Fluorescence emission spectra of isolated collagen fibers measured with a CAM: purified type I fibers, 1;
purified type III fibers, 2; native fibers from a tendon, 3; and native fibers of intramuscular connective
tissue, 4.

degree of separation. Note that the two strong spectra (1 and 3) show an increase
from 400 to 410 nm while the two weak spectra (2 and 4) show a decrease. The
problem to be answered concerns the shape of the spectra. Although all the spectra
were measured in the same way and spectra 1 and 3 appear reasonable, do spectra
2 and 4 really have a maximum at 400 nm?

The fluorescence blank is a measurement of the background fluorescence in the
absence of sample. There is generally a risk of keratin dust being present somewhere
in an optical pathway, and also there may be stray fluorescence from sources such
as the mounting medium. Similarly, optical fibers may exhibit intrinsic fluorescence
or, with a bifurcated probe, there may be cross-talk between outgoing and ingoing
fibers (Chapter 11). Thus, in programming, it is reasonable to measure an appropriate fluorescence blank at each wavelength, to store these data as a vector, and to
subtract them from the vector obtained by measuring a sample. It might also be
appropriate to have a variety of blanks for different types of tissue autofluorescence,
against which fluorescent dye-binding could be evaluated (Andrejevic et al., 1996).

Unfortunately, the excitation maximum for connective tissues is around 370 nm
which is fairly close to the emission peak at 410 to 430 nm. For an ordinary
fluorescence LM, with an excitation filter and a dichroic mirror rather than an
adjustable monochromator, problems are caused by the reflectance, from the sample,
of the upper limit of the excitation bandpass. For both UG1 and UG5 excitation
filters the upper limit exceeds 400 nm. In samples with a low fluorescence intensity
(such as spectra 2 and 4 in Figure 8.9), reflectance from the sample of the upper limit
of the excitation bandpass might occur at 400 nm. By deduction, this would imply
that the corresponding segments of spectra of samples with a high fluorescence
intensity also are unreliable and that their shape is determined as much by the
apparatus as by the fluorescence of the sample.

At first sight, it might appear that an empty-area measurement (corresponding
to a cuvette with solvent but no solute in conventional fluorometry) would be a

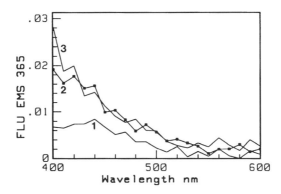

FIGURE 8.10

Pseudofluorescence caused by reflectance, from aluminum foil, of the upper edge of the excitation bandpass. The bandpass of the excitation monochromator was increased from 10 nm (1) to 20 nm (2) and then to 28 nm (3).

suitable fluorescence blank for LM fluorometry; however, even if a sample is not fluorescent, it may reflect the upper edge of the excitation bandpass and this may be mistaken for fluorescence. This may be demonstrated by mounting a clean slip of aluminum foil in distilled water between a slide and a cover slip, ringed with a sealant to prevent evaporation. With the aluminum foil to one side of the light path, the photometer is standardized by the classical method using transmitted light from a halogen source, and the standardization data are adjusted for the emission spectrum of the source. A typical epi-fluorescence condenser contains a heat-absorbing filter, a red-attenuation filter, an excitation filter, a dichroic mirror, and a barrier filter. This configuration was altered by replacing excitation and barrier filters with two grating monochromators. Excitation was from a 100-W mercury source with a quartz collector. The fluorescence blank first was measured with a 10-nm bandpass on the excitation monochromator, then pseudofluorescence (reflectance) was measured from the aluminum foil (Figure 8.10). When the bandpass of the excitation monochromator was increased to 20 nm and to its maximum of 28 nm, pseudofluorescence at 400 nm was approximately doubled. Measurements at 600 nm (well beyond the bandpass cutoff) were virtually unchanged. Thus, when a fluorescence blank has negligible reflectance, then reflectance from the sample of the upper edge of the excitation bandpass may appear as pseudofluorescence. For the subjective observation of fluorescence, the manufacturer of the epi-fluorescence condenser corrects for this effect with the barrier filter, at least as far as is possible (Pluta, 1989). For fluorometry, however, one must chose between the distortion of pseudofluorescence or the distortion caused by a barrier filter. A reasonable solution to this dilemma is to make the fluorescence blank a nonfluorescent surface, such as dull aluminum foil, with a reflectance similar to that of the sample. Aluminum foil is particularly useful because its properties are unchanged by magnification, and it may be used to harmonize microscopic, semi-microscopic, and macroscopic fluorometry.

Chapter  9

Video

9.1   Introduction

Video image analysis (VIA) is the most widely used aspect of computer-assisted light microscopy. Excellent textbooks on the subject are available (Russ, 1990), and many readers already may have access to on-line information for their own specific commercial systems. A video camera and PC frame grabber board can be exploited as a versatile photometer system for a variety of experiments, such as following changes in blood vessel diameters (Fischer et al., 1996) or following cell movements (Schinagl et al., 1996). But with video images of moving subjects or three-dimensional structures, image retrieval and reconstruction become computationally complex (Thomas et al., 1996; Ong et al., 1996; Chen et al., 1996) and are beyond the scope of this book, for which the main subject is direct computer operation or control of the LM. However, low-level elementary VIA principles are useful in programming a CAM, as in the analysis of data collected with a scanning stage.

When a video camera is located on an ordinary LM, it usually takes the place of a photographic camera on a vertical tube, replacing the camera mount with a C-mount for which most video cameras are threaded. On a CAM, where a PMT and monochromator usually have the prime location in the vertical axis of the microscope, probably it is best to locate the video camera as an alternate optical pathway, bypassing the monochromator. The main problem is that VIA ties up the whole of the PC, requiring a large RAM space to hold full-screen images. Although the CAM may be operated from another window of the PC, it is difficult to run both VIA and the CAM at the same time with the same program. The easiest solution is to dedicate a second PC to running the frame-grabber board, uploading the required parts of images to the primary controller.

Video image analysis may be used with polarized light, as in the quantitative determination of mineral fibers (Lundgren et al., 1996). Changes in the brightness of birefringent structures depending on their orientation with respect to the crossed polarizers may be a problem. This may be avoided with circularly polarized light. Circularly polarized light may be created with a quarter-wave plate at 45° to each plane polarizer, one above and one below the specimen (Frohlich, 1986; Pickering et al., 1996). Thus, regardless of how the specimen is rotated or in what pattern the birefringent structures are located, the brightness remains equal.

For certain operations involving a high degree of user knowledge, such as the delineation of structures to be measured, there may be no need to use a video camera. Instead, the CAM may be fitted with a camera lucida or drawing attachment. This device allows an image of the operator's hand to be superimposed on the field of view. Thus, with a light-emitting diode on the tip of a digitizer pen or the center of a mouse, the drawing attachment enables immediate digitization in the field of view (Heilbronner, 1988; Mercer et al., 1990). For those who prefer to work dynamically with an original image, with easy access to the fine focus of the LM, this may be preferable to working on a static video image on a screen.

9.2   Pixels and Gray Levels

The basic components required are a live video signal from the video camera mounted on the CAM and a frame-grabber board in the secondary PC. An inexpensive frame grabber giving a relatively low resolution 512×512-pixel array is more than adequate for our purposes here, where only a 50×64-pixel area of interest (AOI) is passed to the primary controller. Low-cost boards are quite adequate for the CAM, because processing speed and frame storage requirements are relatively modest (Nys et al., 1991). The required image is fitted to the AOI by the choice of objective magnification. It is important to match the width:height aspect ratio of the video camera (such as 4:3) to the aspect ratio of screen and printer pixels, in order to maintain isotropic scaling (so that a circle does not become an ellipse when displayed or plotted).

With an 8-bit ADC, each pixel can be set at one of 256 levels of light intensity (gray levels) from 0 to 255, although the lowest few usually contain electronic noise and may not be much use for analysis. A range of 256 gray levels (which exceeds that of the human eye) is adequate for mapping structures but may not be adequate for absorbance measurements (because of the logarithmic nature of absorbance). This may explain unsatisfactory results obtained with a low-performance video camera being used instead of a PMT for spectrophotometry.

9.3   Video Cameras

Two basic types of video cameras are available, the older vidicon tubes and the newer CCD (charge-coupled device) camera. The former is more likely to be available from recycling than the latter, which is why we need to consider some of its properties. In a vidicon tube, an image is focused on a small window at the end of the tube where the light reacts with a photosensitive conductive coating on the inner side of the window. The window is continuously scanned in a raster or comb pattern by an electron beam from the cathode at the other end of the tube. Thus, the image first is represented on the photoconductive coating as a pattern of high and low resistance regions and then is converted to a current flow by the scanning beam. All this happens at high speed, to give 30 frames per second, so that the main function of the frame-grabber is to latch this fast signal and make it available to the relatively slow secondary PC.

Dust is always a potential problem with a CAM, but the problem gets really serious with video cameras. Never strip down a video camera in an open laboratory, because dust may be drawn electrostatically to the vidicon window and it will appear in the image plane when the camera is replaced on the CAM. If a laminar-flow hood is not available, try a walk-in cooler. Allow the camera several hours to equilibrate before opening it, and allow an overnight warm-up period before using it again (because of condensation on the circuitry). Anti-static cleaning equipment from the photographic darkroom may be helpful.

Users of older, recycled components are likely to acquire first-hand experience with classic vidicon problems. If the video camera is used to record bright spots on a dark-field, the corresponding highly activated spots on the inner photoconductive layer may start to bloom or expand outwards. Thus, a bright area may bloom to fill a large part of the screen, giving a positive bias to morphometry of the enclosed area by pixel counting. Another problem may originate from the deflection and focusing coils of the vidicon camera failing to make a perfect rectilinear scan on the photo-sensitive coating. Thus, the edges of the frame may be pushed inwards (pincushion effect) or may bulge outwards (barrel effect). Similarly, the edges of the window may have thicker glass than the center of the window, so that the image is dim at the edge. If the overall brightness of the image is changed, the proportionality between the output signal and the brightness of measured spots may change across the whole picture. Similarly, it may be wise to avoid the automatic gain option available on a camera. It adjusts automatically to different overall levels of brightness, and a new setting may have a different proportionality between brightness and the output voltage to that which existed in the previous setting.

The CCD camera uses a chip with an array of photodetectors, rather like a photodiode array (Chapter 10). Photoelectrons are trapped in potential wells until read and reset by scanning across the rows and down the columns, which produces an analog voltage representation of the image. An interline-transfer CCD has a parallel buffer for each photosensor connected via a charge transfer gate to enable the buffer registers to be read. The read-out may be of the whole array when in field integration mode, or interlaced half-frames when using alternate rows or image lines to re-assemble an image with a minimum of flicker. Differences in detector performance within the array on the chip may appear as noise in the video signal. While the spectral sensitivities of a vidicon tube can be matched to those of the eye, this is difficult with CCD cameras, which tend to be specially sensitive to red and infrared light. Thus, out-of-focus infrared light may fog the CCD image. Usually, this is prevented by an infrared filter (but this may be removed or replaced with a visible-light filter to obtain infrared imaging).

An inexpensive solid-state color camera may suffer a threefold loss of resolution if it is based on a monochrome chip with three filters for the essential colors — red, green, and blue (RGB). At higher levels of performance, the three basic possibilities are a three-chip camera (one chip for each primary color, giving high resolution), a stripe-filter camera (with separate vertical columns for R, G, and B, and a loss of horizontal resolution), and a mosaic camera (groups of four adjacent photosensors with complementary colors cyan, yellow, green, and magenta, giving an intermediate performance between three-chip and stripe-filters, but not true RGB). Standardization of RGB signals against black and white standards is important experimentally because the emission spectra of illuminators may change with the age of the source, as may the sensitivity of the camera. Thus, if the system is not properly standardized, data cannot be compared from one set of measurements to another. True-color CCD cameras have many advantages for measuring fluorescence intensity ratios, replacing monochrome CCD cameras used in conjunction with filter changers in older microscopes (Bornfleth et al., 1996). Gamma correction is when the output from the CCD camera is corrected for photometric nonlinearities between light in and signal out. Procedures for the evaluation of video cameras on the LM are given by Tsay et al. (1990).

9.4   Software

9.4.1   Noise

Noise may appear as white speckles in a video image. But human vision is amazing in its ability to ignore noise and grasp the main features of an image, so that even video images that look reasonable may contain an impossible amount of noise that overwhelms an attempt at direct computational analysis. To help illustrate techniques for dealing with noise, the following figures were obtained with a frame-grabber giving 256 gray levels for each pixel. Using a small 50×64-pixel AOI gave individual pixels large enough to be seen separately in the figures. To create black and white diagrams, the gray level of each pixel was converted to black and white stippling, with the relative numbers of dots in the stippling pattern giving the gray level. A form of contrast enhancement already has been done, spreading the gray levels over the whole dynamic range of the printer stippling patterns.

9.4.2   Contrast Enhancement

A few words on contrast enhancement are appropriate. A standard video technique readily adaptable to many CAM experiments is to pool all the pixels, regardless of their position in the image, and then to plot them as a histogram showing the number of pixels at each gray level. For example, in a monochrome image composed of

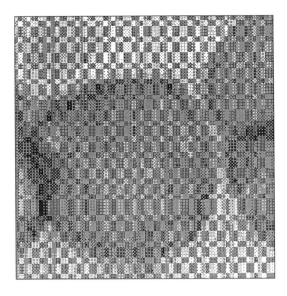

FIGURE 9.1

An adipose cell stained with methylene blue. In this and following figures, to enhance the visibility of effects, fake large pixels were created computationally by integration. Each pixel is a tall rectangular box, filled with a LaserJet stippling pattern to indicate the gray-level intensity of the pixel.

closely related gray levels, the pixels will be grouped over a relatively small fraction of the range from 0 to 255, with none completely black (gray level 0) or white (gray level 255). By transforming the existing gray levels mathematically (subtracting the minimum gray level from the lower gray levels and adding to the higher gray levels), they can be spread over the whole dynamic range from 0 to 255. Then, when returned to their original positions in the image, the contrast will be greatly enhanced so that there are black shadows and white highlights.

With a color image represented by three separate matrices for each of R, B, and G, another possibility is to plot the pixels by their R vs. G coordinates (regardless of their original position in the image). Different clusters may be associated with differently stained tissue components. If they are completely separate, a diagonal across the graph separating the two clusters may be found very rapidly, but if the clusters overlap, a statistical cluster analysis technique may be required. Once this has been done, however, each pixel may be given a value of 1 if it is from a wanted structure, or a value of 0 if from an unwanted structure. These values then are returned to the image matrix using the original x,y coordinates to create a maximum contrast monochrome image. The binary numbers in such a matrix make efficient use of memory and may be processed rapidly using Boolean algebra. For example, an edge-finding algorithm may be used to move around the perimeter of a wanted area, simply by moving ahead while keeping one edge of the leading point in contact with the area.

Figure 9.1 shows an area about 100×100 μm on the sample. The dominant round structure is an adipose cell stained with methylene blue. Parts of other cells are

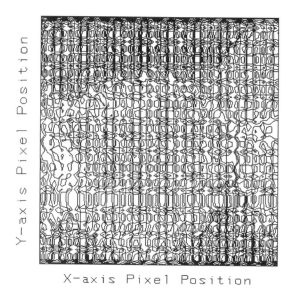

FIGURE 9.2

Failure of a contour plotting algorithm on the raw data of Figure 9.1.

nearby. Plotting the data with enlarged pixels makes noise problems easier to see, but the same principles also apply at higher resolution with small pixels. Contour lines may be drawn around topographical features of a landscape at convenient separations of altitude, and the same technique is useful for VIA, using gray levels in place of altitudes. Figure 9.2 shows a contour map of Figure 9.1, demonstrating that there is so much noise in Figure 9.1, that direct analysis of the raw data is not likely to yield much of interest. Similarly, a three-dimensional view of the data (Figure 9.3) reveals nothing of the cell visible subjectively in Figure 9.1. The solution to this problem is to smooth the image to remove most of the noise. Contrast-enhancement may be useful in combination with differential interference-contrast when examining cultured cells (Tomita et al., 1995).

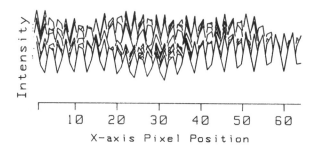

FIGURE 9.3

Failure of a three-dimensional view of the raw data in Figure 9.1.

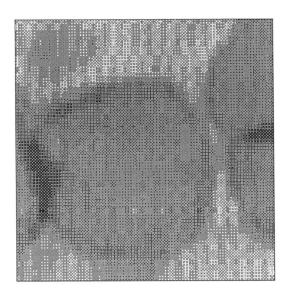

FIGURE 9.4
Smoothing of Figure 9.1 with a simple neighborhood averaging kernel.

9.4.3   Neighborhood Averaging Kernels

A video matrix may be smoothed with a kernel — a small matrix moved systematically over the image matrix, transforming the image by weightings specified in the kernel. For example, each pixel of the image may be averaged with the four adjacent pixels with which it shares a side. Even this very simple method dramatically improves Figure 9.1 to give Figure 9.4. This simple method of neighborhood averaging has a kernel,

$$0 \quad 1 \quad 0$$

$$1 \quad 1 \quad 1$$

$$0 \quad 1 \quad 0$$

where the center point is the pixel being averaged with its four immediate neighbors: above, below, left, and right of center. Corner positions are unused (0 weighting), and the others have equal weighting (1). The kernel moves across the image matrix so that each pixel, in turn, lies at the center of the kernel. The value at the center point is replaced by the average of the five pixels set to 1 in the kernel. However, on the outermost rows and columns of an image matrix, not all the kernel positions are available for averaging, so less pixels are averaged than for internal rows and columns. Working row by row down an image matrix it is important to use pixel intensities from the original or preceding image, and pixels that already have been averaged should not be re-used. Thus, to re-use the existing matrix to store the newly processed image, a few extra rows (the height of the kernel) must be held in a memory buffer to avoid averaging data that have already been averaged, unless a recursive filter is being applied (Zimmer, 1979).

The weighting of the center pixel may be increased to reduce the chance of losing details:

$$
\begin{array}{ccc}
1 & 2 & 1 \\
2 & 4 & 2 \\
1 & 2 & 1
\end{array}
$$

where the center pixel has a weighting of 4, the major neighbors 2, and the farthest neighbors 1. The weighting of the center pixel may be increased for even greater preservation of a single pixel detail, but then it might be better to increase the size of the whole kernel to maintain the averaging effect of canceling noise between pixels, such as:

$$
\begin{array}{ccccc}
1 & 2 & 3 & 2 & 1 \\
2 & 7 & 11 & 7 & 2 \\
3 & 11 & 17 & 11 & 3 \\
2 & 7 & 11 & 7 & 2 \\
1 & 2 & 3 & 2 & 1
\end{array}
$$

Kernels much larger than those shown above are used in commercial software, and the weighting of the center point may be Gaussian. With a median filter, instead of replacing the center pixel by the arithmetic mean, the pixels of the kernel are ranked in order, then the median value is used to replace the center pixel. Sometimes median filters may be better than averaging filters because they maintain the brightness differences between pixels and cause less move-ment of boundaries. Three passes of a median filter were used to smooth the raw data of Figure 9.1 to produce the image shown in Figure 9.5. Many a highly improved image has been discarded by the operator because it looks worse than the original, but before dumping Figure 9.5, let us see how it yields to contour (Figure 9.6) and three-dimensional analysis (Figure 9.7). The result is a dra-matic improvement which could be searched computationally for its main fea-tures. If spots and small areas are the main subject of measurement, then a top-hat filter may be used to preserve information at the center of a kernel (Bright and Steel, 1987).

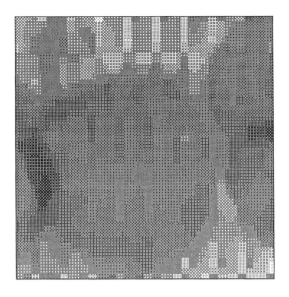

FIGURE 9.5
Three passes of a median filter over the raw data of Figure 9.1.

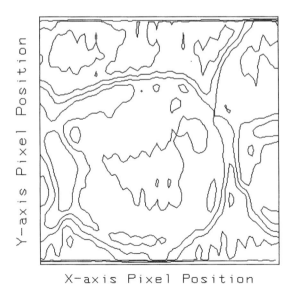

FIGURE 9.6
Contour plotting of the smoothed data of Figure 9.5.

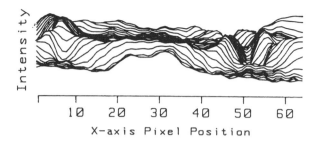

FIGURE 9.7
A three-dimensional view of the smoothed data of Figure 9.5.

9.4.4   Laplacian Kernels

Kernels may be adapted for sharpening boundaries or increasing the contrast between structures in an image, as in a Laplacian kernel.

$$
\begin{array}{ccc}
-1 & -1 & -1 \\
-1 & +8 & -1 \\
-1 & -1 & -1
\end{array}
$$

Moving the kernel across relatively smooth areas produces little effect, but a strong effect occurs at boundaries, giving sharper images for the human eye. However, this is only because the eye is very responsive to boundaries and there may be no improvement for computational analysis. Averaging kernels tend to act as low-pass filters, letting through broad shapes but filtering out noise, while Laplacian kernels tend to act as high-pass filters, missing broad shapes but emphasizing regional differences, which are easily influenced by noise.

9.4.5   Erosion and Dilation

From the data of Figure 9.5, pixels < 128 were converted to 0 and the remainder to 1, producing the binary image shown in Figure 9.8. When two intrinsically separate areas touch each other to produce an apparently continuous structure, sometimes they can be separated by erosion and dilation. In the erosion operation, pixels are turned off (set from 1 to 0) all around the borders of the major areas, thus reducing the area. After several erosions, a crack separating two adjacent areas that touch may be completely opened. Once this is detected, the area is dilated by turning on the pixels around its edges, except where this would cause rejoining of two separate areas. Dilation is repeated until the areas are back to their original size, except in the seam where adjacent areas would touch. In opening seams between contiguous areas it is useful to know the likely radius of curvature of the separation, found from a nearby

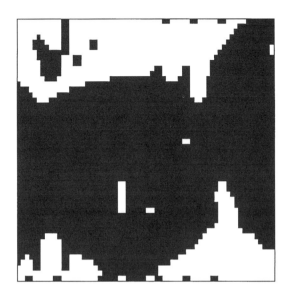

FIGURE 9.8
A binary image of the smoothed data of Figure 9.5.

exposed region, or to jump along a series of separated pixels in the general direction of the separation.

Erosion and dilation operations are available in most general-purpose software packages for VIA. In a typical application, the operator writes a macro that goes through a series of erosions and dilations for the most difficult separation in the set of images to be processed, then the remainder of the images are processed automatically. But there is still considerable subjectivity in the operation, and a macro that works satisfactorily for an ideal image may fail for a typical image. Interactive decision-making by the operator is almost essential for erosion and dilation methods, and complete automation of erosion and dilation usually is unreliable.

9.5   Summary

In using a video camera for CAM experiments, digital filtering may be required before an image can be used computationally. A high level of VIA is difficult to sustain in the same operating environment as that used for CAM programming, but applications may be found in gathering morphometric information and in searching for specific simple structures, such as isolated cells in a smear using a scanning stage.

Chapter **10**

Photodiode Array Spectrography

10.1 Introduction

A photodiode array (PDA) consists of a line of up to 1024 photodiodes mounted on an integrated circuit. Each photodiode consists of a p-zone inset into the top of a continuous n-zone. Parallel PDAs exist, with multiple outputs, but self-scanning PDAs are of more use in the CAM. As in the PIN diode (Chapter 4), exposure of each photodiode generates a current. The photovoltaic current discharges a capacitor, and the extent of discharge is determined when the photodiode is read and refreshed. Photodiodes are scanned in sequence, and the output is fed to the video line of the PDA. Thus, in a PDA spectrograph, where a spectrum is dispersed across the PDA by a prism or diffraction grating, the whole spectrum can be read in 5 to 100 msec. If the time needed to measure a spectrum is critical, a PDA spectrograph may provide an ideal solution, provided that the light intensity is relatively high.

10.1.1 Advantage of PDA Over Shutter Pulse

As an example, consider the problems posed by measuring a UV fluorescence emission spectrum with a motorized spectrophotometer. If the fluorescence quenching is very rapid, then the last measurements of a spectrum are negatively biased relative to the first measurements. In other words, by the time the spectrophotometer reaches the last wavelength to be scanned, the fluorescence has faded. This problem can be reduced by pulsing the UV with a shutter and using the rising slope of a photomultiplier response (Chapter 3), but a PDA spectrograph to measure all wavelengths almost simultaneously may be a better solution.

Using a protective shutter (Chapter 3) and a motorized spectrophotometer, it is difficult to measure the whole visible spectrum with less than 3 sec of exposure of the sample to UV (31 wavelengths × 0.1 sec shutter open time). However, the same information may be acquired with only a 0.5-sec exposure of the sample using a PDA. This pulse of UV contains ample safety margins for opening a relatively slow, swing-in shutter and for operating the PDA and, probably, could be reduced still further if necessary.

10.2   Mounting the PDA

Mounting a PDA spectrograph directly on a CAM so that the optical axes of the two components are properly aligned calls for some expensive, precision engineering. At extremely low light intensities, a professionally engineered CCD is required, which is beyond the scope of this book. However, with high light intensities made possible by using a relatively large photometric aperture and/or medium to low magnification, it is possible to link the optical axis of the microscope to the PDA spectrograph using a fiberoptic light guide. There are coupling losses, particularly from light lost into the cladding and gaps between optical fibers. But, if necessary, these can be offset by cooling the PDA to –15°C for integration times of up to 10 sec.

An overview of a PDA connected to a CAM is shown in Figure 10.1. The main controller was an HP BASIC workstation linked via the GPIB to a slave controller dedicated to running the PDA (a 486 PC with National Instruments 488.2 GPIB). The PDA was potentially capable of running at 25 Hz and required a large RAM buffer for real-time studies. However, to keep this description simple, we will only deal with taking one spectrum a time. For real-time studies, time simply becomes the secondary scanner, with the PDA spectrograph as the primary scanner.

10.2.1   PDA Spectrograph

The PDA was an EG&G Reticon S series (Sunnyvale, CA) with 1024 elements mounted on an asymmetric Czerny-Turner configuration monochromator (model FF8810, Sciencetech; London, Ontario), shown as λ in Figure 10.1. The PDA was operated from a Sciencetech LDA 2000 scanning control unit, using a customized Advancetech 30 kHz 12-bit ADC. The monochromator was linked via two back-to-back achromats (f = 50 and 75 mm) and a light guide (5-mm diameter) to the point where a side-window photomultiplier is normally mounted in a Zeiss type 01K photometer head. This photometer head has a user-rotated mirror directing light either to the photometer or to the eyepiece, so that observations and measurements cannot be made simultaneously. Thus, tissues were identified and positioned by oblique illumination with visible light, then the mirror was rotated to allow measurement of the target tissue during fluorescence excitation with short pulses of UV. This allowed full intensity to be directed to the PDA (instead of being split between the eyepiece and PDA).

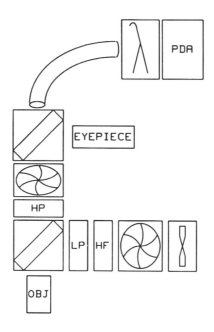

FIGURE 10.1

Block diagram of PDA and Czerny-Turner monochromator (λ) connected via a light guide to the optical axis of a Zeiss Standard microscope, with a 01K photometer head and eyepiece, and a IV/F epi-fluorescence condenser (high-pass filter, HP; objective, OBJ; low-pass filter, LP; and heat filter, HF).

10.2.2 CAM

Ultraviolet excitation was from a 100-W mercury arc with a stabilized power supply (Zeiss 482590). A solid swing-in protective shutter (Zeiss 467226) was mounted in front of the arc and was opened for 0.52 sec per measurement. A faster, segmented shutter could have been used, but at the risk of heat damage from the arc lamp. A delay of 0.2 sec was programmed to occur after the instruction to open the shutter, thus allowing time for the shutter to open and for the PDA to cycle before latching one output cycle by an external trigger. The external trigger to read the PDA used coaxial cabling and the pull-up voltage from the source card.

A Zeiss epi-fluorescence condenser IV/F on a Zeiss Standard microscope was used with a heat filter (HF in Figure 10.1), UG-1 low-pass filter (LP in Figure 10.1), dichroic beam splitter at 395 nm, and high-pass filter (HP in Figure 10.1) at >420 nm. The microscope objectives were Zeiss neofluars.

10.3   Retrospective Standardization

The optical axis of a Czerny-Turner monochromator has an M-shaped configuration, with the input slit and input mirror on the left, the exit mirror and PDA on the right,

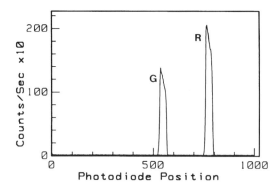

FIGURE 10.2
Standardization signals from green (G; 543.5 nm) and red (R; 632.8 nm) helium-neon lasers.

and the grating in the midline. Thus, rotation of the grating changes the spectral dispersion across the PDA for different applications. To standardize the spectrograph at any particular grating position, the microscope objective was focused on the cut, polished end of an optical fiber mounted vertically in the stage. The other end of the optical fiber was connected to an optical bench with a rotatable mirror that allowed light from either a green (543.5 nm) or a red (632.8 nm) helium-neon laser to be directed into the optical fiber. Side-mode dispersion of the laser output spectrum was ignored. Thus, the dispersion of essentially monochromatic light from the laser provides a measure of the monochromator bandpass, as well as the primary standardization wavelengths. If lasers are not available, the peaks of the mercury arc emission spectrum may be used for standardization, but this requires careful programming to identify the correct mercury lines if the monochromator is at a completely unknown setting.

The photodiode positions along the PDA corresponding to the two lasers were found (Figure 10.2). The relative positions of monochromatic light from the two lasers enabled the photodiode position along the PDA to be converted to wavelength, while the widths of the peaks revealed the effective band-width of the spectrograph. Peak height was unimportant (because it was determined by the coupling efficiency of the optical fiber to the source laser, the nature of the optical fiber, and laser intensity).

This procedure facilitated finding an appropriate balance between resolution and transmittance of the monochromator. Thus, the optical system was manipulated manually until suitable spectra were obtained from a sample, then the optical system was standardized retrospectively, for which purpose the optical fiber carrying the calibration wavelengths from the lasers was mounted in a hole drilled into the stage, just to one side of the specimen.

The PDA output (counts per sec) was converted to relative fluorescence intensity (Chapter 8). A halogen source with a known emission spectrum was used to find the spectral response of the PDA, which then was used to generate correction factors for each wavelength (at each photodiode position along the spectrum dispersed across the PDA). Thus, had the PDA been exposed to a hypothetical ideal light source with a flat emission spectrum, all photodiodes along the PDA should have given the same response. This was simpler than a theoretical protocol, working in the opposite direction and starting with the manufacturer's data on the spectral response of the PDA, because it would have been necessary to correct for monochromator efficiency and transmission characteristics of the light guide linking the spectrograph to the microscope. The system was tested by measuring the fluorescence emission spectrum of GG17 uranyl glass, and the error was considered acceptable (the fluorescence peak was found at 536 nm instead of 530 nm).

Factors converting photodiode position to wavelength can be maintained in the array holding photometer variables (Chapter 4) once they are known.

{1} = Minimum wavelength of PDA

{2} = Operative number of photodiodes

{3} = Nanometers per photodiode

{4} = Trigger mode (software or external trigger)

{5} = Lowest wavelength to show in graphics

{6} = Highest wavelength to show in graphics

However, the operator then must be prevented from altering the monochromator settings. The operative number of photodiodes may be less than the hardware maximum (1024) because of latching problems. Having irregular minimum and maximum values, the spectrum from the PDA may need clip limits for neat plotting.

10.4 Fluorometry of Collagen and Elastin

Bundles of collagen and elastin fibers were trimmed to small blocks (each with a volume of several cubic millimeters) and stored in 40 ml 0.2-M phosphate buffer (six subsamples from pH 5 to pH 7.5 in steps of pH 0.5 for 7 to 10 days at approximately $2°C$). Sample blocks were cut into bundles of fibers under a dissecting microscope (some were cut into disks across the fibers to be measured down the longitudinal axes of fibers, and some were cut into lengths of fibers to be measured perpendicularly to the longitudinal axis of fibers). Samples were mounted temporarily in buffer solution on microscope slides immediately prior to measurement.

10.4.1 Sources of Variance: Samples vs. Photodiodes

Figure 10.3 shows the general features of a set of fluorescence emission spectra measured with the PDA spectrograph. The line shows the mean of the 20 measurements

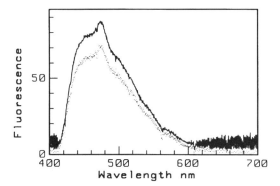

FIGURE 10.3

A set of fluorescence emission spectra of elastin fibers from bovine ligamentum nuchae at pH 5.0, 7 days postmortem, measured at 24°C in a direction perpendicular to the longitudinal axes of fibers. The line shows the mean, and the dots show one standard deviation subtracted from the mean.

made to encompass biological variation along and between samples of the ligamentum nuchae. The degree of sample variation may be seen from the dots, which show one standard deviation subtracted from the mean at each photodiode along the PDA. Thus, by chance, at a few wavelengths where no fluorescence was detected, the mean minus one standard deviation was less than zero. Single spectra exhibited the same features as those shown by the mean in Figure 10.3. There was relatively high variation between adjacent photodiodes along unilluminated lengths of the PDA (for example, at photodiodes corresponding to 620 to 700 nm) which generated a sawtooth pattern. This was not smoothed by averaging the spectra from different sample sites, which showed that it originated from the PDA spectrograph. For illuminated photodiodes corresponding to the wavelengths of the fluorescence emission spectrum, the sawtooth pattern was reduced or absent in both single spectra as well as in the mean.

From the corresponding standardization data (Figure 10.2), it may be seen that monochromatic light was spread over approximately 50 photodiodes at 0.393 nm per photodiode. Thus, the intensity of PDA dark-field noise and inter-photodiode variation along the PDA (for example, from 620 to 700 nm) could have been reduced relative to the height of the fluorescence peak, but this would have caused a corresponding loss of spectral resolution.

A convenient method to smooth the data (when appropriate) was to convert each spectrum to a histogram, using histogram columns for one experimental treatment and marking the tops of columns with a + for a comparison treatment. For example, Figure 10.4 shows measurements of elastin fibers made perpendicularly to the long axes of fibers at pH 5 (columns) vs. pH 7.5 (+). With 20 measurements of each group, the separation of the means at different pH values was significant, $P < 0.005$.

As introduced in Chapter 1, another method of analysis was to plot the results of appropriate statistical tests by wavelength, as in Figure 10.5, which shows the distribution of the t-statistic for a comparison of the fluorescence of collagen fibers

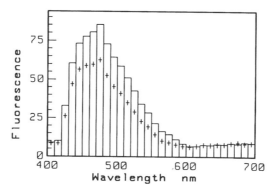

FIGURE 10.4

Histogram analysis of the fluorescence of elastin fibers at pH 5 (columns) vs. pH 7.5 (+) for measurements made perpendicularly to the long axes of fibers at 7 days postmortem at 24°C (n = 20 for each type).

at pH 7.5 vs. pH 5.0. Measurements were made perpendicularly to the long axes of fibers at 7 days postmortem at 24°C (n = 20 for each type). Although the fluorescence formed a peak as in Figures 10.3 and 10.4, the t-statistic formed a plateau across almost the whole width of the peak. However, across the fluorescence peak (from 430 to 520 nm) there was a large variation in the t-statistic (t ≈ 5) among photodiodes along the PDA. Although all were significant (P < 0.005) in this experiment, this would not always be the case. It would be prudent to use some method of smoothing or averaging across the spectrum in such situations to avoid type I statistical error (rejecting the null hypothesis when true).

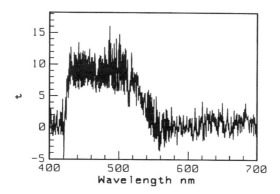

FIGURE 10.5

Spectral distribution of the t-statistic for a comparison of the fluorescence of collagen fibers at pH 7.5 vs. pH 5.0 for measurements made perpendicularly to the long axes of fibers at 7 days postmortem at 24°C (n = 20 for each type).

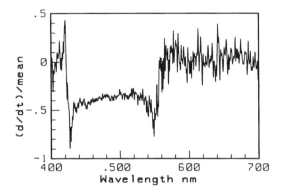

FIGURE 10.6

Relative rate of fluorescence quenching in elastin fibers during 2.1-min exposure to UV, measured down the longitudinal axes of fibers at 24°C 12 days postmortem at pH 5.0 (data smoothed with two passes of a moving average).

10.4.2   Software for Rate of Quenching

By changing the programming instructions for the shutter protecting the sample from UV it was possible to monitor the rate of fluorescence quenching. The shutter was opened immediately before the first measurement, left open between measurements, and shut after the last measurement. As would be expected, the largest change occurred at the fluorescence peak, so that the absolute values for the decrease in fluorescence with time (dF/dt) mirrored the basic shape of the fluorescence emission spectrum. This obscured the true nature of relative changes across the spectrum. This problem was solved by using (dF/dt)/mean, where the mean value was the mean fluorescence across the time series. For example, Figure 10.6 shows the relative quenching rate for elastin fibers at pH 5.0. On a relative basis, the greatest rate of quenching was on each edge of the peak, not at the peak. However, the statistical significance of the change was minimal in these regions (where relatively small changes in fluorescence overlapped dark-field noise and inter-photodiode variation along the PDA), and the distribution of the t-statistic mirrored the fluorescence peak (Figure 10.7).

10.4.3   Effect of pH and Measuring Direction

Biological specimens are notoriously variable in their properties and the following observations may not be universally valid, but in the tissues used for this example the fluorescence of elastin fibers was always stronger ($P < 0.005$) than that of collagen fibers (for example, compare Figures 10.4 and 10.8), and fluorescence was always stronger at a low pH (pH 5 or 5.5) than at a high pH (pH 7.0 or 7.5), as shown in Figure 10.4. All samples showed a trend towards stronger fluorescence when measured

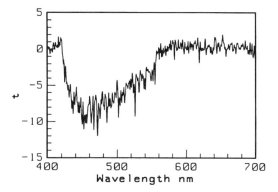

FIGURE 10.7

Distribution of the t-statistic for data in Figure 10.6 (for null hypothesis slope = 0, n = 10 measurements, and data smoothed with two passes of a moving average).

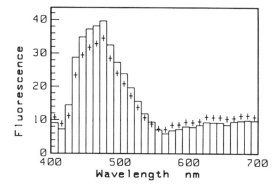

FIGURE 10.8

Histogram analysis of the fluorescence of collagen fibers at pH 5 measured down the longitudinal axes of fibers (columns) vs. perpendicularly to the longitudinal axes of fibers (+), 7 days postmortem at 24°C (n = 20 for each type).

down the longitudinal axes of fibers rather than perpendicularly across the fibers, but this was only significant (t > 4, P < 0.005, $\lambda \approx 425$ nm) for collagen fibers at a low pH (Figure 10.8). Since the depth of tissue in the field was approximately constant, the nature of this possible directional effect and its relative weakness in elastin, despite the stronger fluorescence of elastin, warrant further investigation.

Chapter **11**

Liberating the Microscope with Fiberoptics

11.1   Introduction

Optical fibers are best known for their applications in digital telecommunications, but optical fibers also may be used for the direct transmission of analog information, such as spectra and images (Kapany, 1967). Thus, optical fibers mounted on a CAM enable it to reach remote locations, just like a remote fiberoptic spectrophotometer (Fitch and Gargus, 1985; Ward and Hussey, 1987). And, if a specimen is too large for the LM stage or cannot be mounted without denaturation, the CAM can be brought to the specimen (Leggett, 1997). Fiberoptics are ideal for custom-built apparatus. They are relatively inexpensive, can reach great distances, have a small diameter, are flexible, and resist moderate heat and acidity (Neti et al., 1986).

In some ways, fiberoptics and VIA are competitors, because both may be used for remote sensing. VIA may be best for obtaining images or spectral data through clear air or water, but optical fibers become competitive for internal imaging (endoscopy) and for penetrating inside pipes, enclosures, or turbid media. Heat that might destroy a video camera may produce only a slight, predictable bias on data collected by a quartz optical fiber. There are many ways in which optical fibers can be used experimentally, as in the measurement of pH (Benaim et al., 1986; Alabbas et al., 1996), temperature (Adamovsky and Piltch, 1986), organochlorides (Milanovich et al., 1986), redox states (Mayevsky, 1984), luminescent enzymes (Blum et al., 1988), NIR spectroscopy of living brain tissue (Delpy et al., 1987), and many others (Peterson and Vurek, 1984). The focus of attention in this chapter will be on relatively simple methods that extend normal CAM operations.

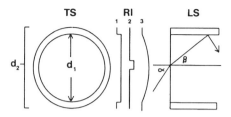

FIGURE 11.1
Transverse section (TS), refractive index (RI) profile 1, and longitudinal section (LS) of a step-index multimode optical fiber. RI profiles 2 and 3, respectively, are for monomode and graded-index fibers.

11.2   Optical Fibers

A light guide is a structure such as a fluid-filled tube or a glass rod with a silvered surface that guides light by a series of internal reflections. Although an individual optical fiber is a light guide, a fiberoptic light guide usually denotes a bundle of optical fibers protected against abrasion and kinking by an outer protective sheath. If the position of an individual fiber within the light guide is constant relative to the other fibers, then the light guide may be used to conduct an image and is called a coherent fiberoptic light guide. Light guides that lack this property are called noncoherent. Textbooks on fiberoptic theory are readily available (Daly, 1984), but some practical information is needed here to understand how optical fibers may be used in conjunction with a CAM.

11.2.1   Step-Index Fibers

Figure 11.1 shows a transverse section (TS), a refractive index profile (RI, 1), and a longitudinal section (LS) of a step-index multimode fiber. The core of the fiber (diameter = d_1, refractive index = n_1) is surrounded by a cladding (thickness = $(d_2 - d_1)/2$, refractive index = n_2) that causes total internal reflection to conduct light along the core. The cladding is analogous to the insulation around an electrical wire in that an optical core still conducts light, provided that the surrounding medium has a lower refractive index. Thus, a primary purpose of the cladding is to prevent short circuits. In Figure 11.1, RI profile 1 shows the step change in refractive index (from n_1 to n_2) where the cladding meets the core of a step-index fiber — hence the name.

11.2.2   Numerical Aperture

In the LS of Figure 11.1, a ray of monochromatic light is shown entering the core at angle α from a medium beyond the fiber with refractive index $< n_1$. The ray is bent towards the normal as it enters the optically denser medium. If angle α is increased to beyond the maximum coupling angle (α_{max}), rays no longer enter the core but are reflected from its surface. The numerical aperture (NA) of the fiber is given by:

$$NA = \sin \alpha_{max} = \sqrt{(n_1^2 - n_2^2)}$$

Rays with angle $\cos \beta < n_2/n_1$ are guided along the core, but different angles of entry will have different path lengths along the core — different modes, hence the name multi-mode fiber. A high NA and a high angle β allow more modes than low NA and β.

11.2.3 Monomode Fibers

Except when using pulses of light or interferometry in optical fibers connected to a CAM (Boppart et al., 1996), multimode dispersion is not a problem (as in digital communication) and step-index fibers are satisfactory. Suitable fibers may range from $d_1:d_2 = 100:140$ µm, used in bundles of about 50 fibers for spectrophotometry, to $d_1:d_2 = 1000:1050$ µm, used for single-fiber fluorometry. Monomode fibers (Figure 11.1, RI profile 2) are essentially step-index fibers with a narrow core. This makes coupling rather difficult but only allows a single mode to propagate, so that mode dispersion is low. In graded index fibers (Figure 11.1, RI profile 3), the refractive index of the core increases parabolically towards the fiber axis so that modes travel in sinusoidal waves. Those with longer light paths through outer parts of the core travel faster than those with shorter paths near the fiber axis, so that mode dispersal is low. The acceptance angle of graded index fibers is high near the axis of the fiber but low towards the cladding boundary. Monomode and graded-index fibers have interesting properties that may provide many future experimental possibilities. For example, Fujii and Yamazaki (1990) used a pair of monomode fibers to form a Michelson interferometer. With the phase-sensing signal applied to one arm of the interferometer, a six-port fiber configuration allowed three-dimensional analysis, limited in resolution only by the 9-µm core of the monomode fiber. Polarization-preserving optical fibers are available but, at present, are difficult to use in coupling a CAM to a remote specimen.

11.2.4 Coupling

The coupling of optical fibers to other optical components always requires care and attention. Small movements can produce large changes in coupling efficiency. Also, changes in coupling efficiency may be a function of wavelength because of chromatic aberration. Many types of optical fibers were developed primarily for data communication with monochromatic red or NIR light. Thus, the dependency of refractive index on wavelength is probably the major unseen problem in utilizing optical fibers with the CAM. A judicious standardization protocol may be used to correct for the effect of wavelength on coupling efficiency but, at every conceivable opportunity, this must be tested by proving that a blank sample will return a flat

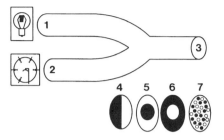

FIGURE 11.2

A bifurcated fiberoptic light guide, with a branch to the CAM illuminator (1) and a branch to the photometer (2). In the common trunk of the light guide (3), optical fibers may be spliced across a diameter (4), with illuminating optical fibers in the periphery (5), or in the axis (6), or with a random arrangement (7).

reflectance spectrum at all wavelengths. This may be more difficult than it first appears, because the reflectance standard must be coupled to the optical fiber with a refractive index medium similar to that which occurs in the fluid phase of the tissue to be measured.

11.2.5   Fiber Arrangements

Light guides may be fashioned into a variety of configurations, the simplest and most important being a splice, where two or more bundles of optical fibers are bound together in the same protective sheath to form a bifurcated light guide (Figure 11.2). At the end of the common trunk, the receiving optical fibers to the spectrophotometer may be arranged in several ways — on one side of a diameter, in the axis, in the periphery, or randomized (Figure 11.2, 4 to 7, respectively). The split diameter arrangement is the easiest to make, and the common trunk may be kept short to limit the cross-talk between receiving and illuminating fibers.

For concentric fiber arrangements, which are difficult to make neatly, a thin layer of metal foil separating axial from peripheral fibers reduces cross-talk and helps maintain fibers in place. Concentric arrangements are suitable for making measurements in liquids, using a mirror at a fixed distance, parallel to the TS of the optical fiber, so that light from axial illuminating fibers is reflected back into peripheral receiving fibers. This enables the optical path length to be fixed. The ratio of receiving to illuminating fibers may be varied to optimize the performance of the system. In most cases, more receiving fibers than illuminating fibers are required.

The randomized fiberoptic pattern is the most difficult to construct and generally requires a long common trunk. True randomness is difficult to achieve, and "scattered" is probably a more accurate description. If no attempt is made to control optical fiber position when a light guide is spliced, the result usually resembles a split diameter pattern.

Different fiberoptic patterns may have different sensitivities to the optical properties of a sample. The general principle is that, to measure absorbance in the liquid phase of a tissue while minimizing any effects from scattering, a relatively long light path must be created to separate illuminating and receiving optical fibers. On the other hand, to measure scattering while minimizing any effects from absorbance, a randomized arrangement of small optical fibers is used because this has the greatest sensitivity to tissues immediately in contact with the optical window.

Various types of optical circuits can be created by twisting single optical fibers together and fusing them with a heat gun or soldering iron. The cores communicate where they connect, and any unused endings are sealed with a light trap. Thus, either receiving or illuminating optical fibers can be split or combined. High-order modes may travel in the cladding, especially after passing through a twisted coupling, but can be removed by passing the naked core of a fiber through a small light trap filled with LM immersion oil.

11.2.6 Internal Reflectance in Tissues

In biological samples, optical complexity tends to be proportional to microstructural complexity. Thus, normal photometric laws may fail, and scattering lengthens the light path, even to the point of it becoming greater than the thickness of the sample (Butler, 1962). In a simple situation, where optical fibers are directed towards, but do not touch, a biological specimen, familiar concepts such as reflectance may be appropriate. Strictly speaking, however, receiving optical fibers in this configuration may be measuring sterance, the collection of light emitted through a solid angle or steradian by a planar area, not reflectance. Conventional transmittance and absorbance measurements are possible if illuminating and receiving fibers are separated by a clear space. But, when illuminating and receiving optical fibers are inserted directly into a translucent tissue with a complex microstructure, the light returned to the spectrophotometer may be a function of many factors, such as:

(1)   Light scattering by particles

(2)   Selective absorbance in the liquid phase

(3)   Reflection and refraction at boundaries where refractive index changes

(4)   Differences between the refractive index of the fiberoptic core and the fluid phase

(5)   Wavelength

(6)   Any optical anisotropy of the sample

(7)   Spatial separation and angular orientation of receiving fibers relative to illuminating fibers

In cases where transmittance and reflectance are inappropriate terms because of these complexities, Conway et al. (1984) suggested the suitably ambiguous term, interactance, to describe the light returned to a fiberoptic system after passing through a complex tissue. Internal reflectance is a reasonable alternative and seems to be more widely used.

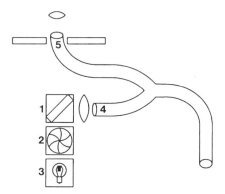

FIGURE 11.3

CAM layout for remote spectrophotometry via a bifurcated light guide. The slide-in mirror (1) isolates the illuminator shutter (2) and illuminator (3) for the illuminating branch of the light guide (4). The receiving branch is mounted in the LM stage (5).

11.3   Remote Spectrophotometry

The easiest way to get started is to remove the illuminator and its shutter, and to focus the collimated output beam onto the illuminating light guide. A better method, which maintains the alignment of the illuminator relative to the CAM, is to use a slide-in mirror immediately after the illuminator shutter (Figure 11.3, 1). The receiving light guide is secured in the LM stage, after first having removed the substage condenser. In the layout shown in Figure 11.3, measurements would be vulnerable to small movements of the light guides close to the LM. Light guide movement near the CAM may be eliminated by shaping supporting conduits out of soft copper tubing. Movement of the distal end or common trunk seldom affects measurements, which allows the common trunk to be used as a probe on remote specimens. However, it is best to secure the common trunk in a clamp with limited movement and, if possible, to bring the sample on an adjustable platform up to the level of the light guide.

When the CAM is configured as a microscope spectrophotometer, the layout in Figure 11.3 allows remote spectrophotometry of a wide variety of specimens, alive or dead, wet or dry, large or small. The light-guide may be as long as is required ($\approx$10 m) to reach an operating table or the experimental apparatus; however, because many types of optical fibers are intended for use with red or infrared light, their transmittance of lower wavelengths may be poor or negligible. Thus, to work at 500 nm or less, it may be necessary to keep the light guide as short as possible ($\approx$0.5 m). Generally speaking, plastic optical fibers are far easier to work with than quartz optical fibers. With practice, they can be shaped and given a superior optical window, working merely with a scalpel under a dissecting LM. Despite what one might first expect, scalpel blades with a sturdy, large angle to their cutting edge provide better results than thin razor blades with an acute cutting angle. Working under hot water, plastic optical fibers can be gently coaxed through surprisingly tight apertures and angles, provided they are internally smooth. The fiber is pulled through as far as necessary to trim off the damaged leading end. Remaining traces of water in the shaft of a probe or hypodermic needle may be removed with ethanol or compressed air. Epoxy resin is used to hold the optical fibers in place, using a vacuum chamber to remove air bubbles. Quartz optical fibers, on the other hand, cannot easily be made to follow a tight radius of curvature and they tend to shatter when cut. Production of an optical surface requires grinding with successively finer grades of carborundum and final polishing. Quartz fibers generally have a higher transmittance of lower wavelengths relative to plastic fibers, although some large-diameter plastic fibers (such as Hewlett-Packard HFBR) have a usable transmittance down to the mercury line at 365 nm. As an added attraction, the HFBR optical fiber is well supported with bulkhead fittings and connectors that can easily be adapted to hold the optical fibers in the LM stage.

11.3.1 Standardization

Standardization of the CAM must include all the optical elements used in making measurements, especially the light guide. It is a cardinal mistake to standardize the CAM, then to add a light guide to make a measurement from a specimen. Because of the variety of possible fiberoptic arrangements in the common trunk of the light guide (Figure 11.2), the common trunk should be clamped in a device that allows precise adjustments in height. At appropriate points in the standardization protocol, the operator is prompted to adjust the height of the light guide above the standard, obtaining the maximum light to the photometer. Thus, the operator needs some appropriate information on the light reaching the photometer. A constantly refreshed number on the screen may be adequate, but a graphics window simulating an analog meter may be easier to read at a distance (if the operator and the distal end of the light guide are both across the room). As always, it is wise to remeasure the standard both before and after data collection.

An absolute white standard for reflectance standardization may be prepared by collecting the smoke from burning magnesium ribbon on an appropriate surface. However, the preparation of consistent reflectance standards in this way is very exacting, and an easier method is to purchase optical-quality barium sulfate powder which already has been calibrated against magnesium oxide. The powder is pressed into a plate with a smooth, regular surface and is used to calibrate a robust tertiary standard such as a plate of opal glass. Preparation of a perfect barium sulfate secondary standard requires sieving the powder and vibrating the press as the powder is compressed.

When the barium sulfate secondary standard is measured via the light guide, it is essential to clamp the light guide closely above the barium sulfate but never to let the light guide actually touch the barium sulphate. The separation is varied until the maximum PMT reading is obtained, so that the operator needs an on-screen display of the PMT output. If the light guide is composed of randomly arranged optical fibers, maximum reflectance will be closer to the barium sulfate than with a split diameter arrangement. A 1-kg batch of optical-quality barium sulfate should last a long time, and its reflectance spectrum relative to magnesium oxide may be embedded in software. The reflectance of barium sulfate relative to magnesium oxide is linear (not a curvilinear spectrum), which enables calibrations to be programmed with a simple linear regression to generate appropriate step increments to correspond with those chosen by the operator for the monochromator. With the CAM standardized against a barium sulfate secondary standard but yielding data as if calibrated against magnesium oxide, the opal glass tertiary standard is now measured.

The distal end of the light guide (away from the CAM, towards the specimen) must be linked to the opal glass with a medium of refractive index similar to that which will be used for measurements of a specimen. If there is to be an air space between the light guide and the specimen, there should be an equal space between the light guide and the opal glass. If the light guide is to be placed in direct contact with wet tissue to make a measurement, then the light guide must be linked to the opal glass by a fluid of refractive index similar to the fluid phase of the tissue. In most cases, a tissue will be protected in an isosmotic buffer, which also can be used for the standardization protocol (it will not be a perfect refractive index match, but close enough optically).

The above procedures are for a simple situation, when the distal window of a light guide in a probe terminates in a simple round window cut perpendicularly to the long axis of the light guide. Elliptical windows cut at an angle to the long axis of the light require more effort, because the light guide now must be tilted at varying angles as well as being held at varying distances above the standards. An even more complex situation may occur if randomized illuminating and receiving optical fibers are splayed and radiated to form an annular window all around the shaft of a probe. This configuration is useful because it has high sensitivity to light scattering in a tissue but low sensitivity to absorbance by chromophores in the liquid phase of the tissue. Primary standardization against optically calibrated barium sulfate powder can be undertaken by rigidly attaching a metal container to the top of a flat-top laboratory vibrator. The probe then may be standardized in the vibrating powder, setting a long integration time on the photometer. If using a PDA spectrograph, it is best to take the highest set of readings from a set of ten. Teflon tape then may be used as a tertiary standard (Weidner and Hsia, 1981), taking care to use enough layers for saturation (10) and not stretching the tape.

Programming for a variety of different standardization protocols rapidly becomes complex. Perhaps the most sensible thing is, at the start of an experiment, to undertake the operations described above, finding the reflectance of the opal glass tertiary standard with the apparatus configured as if to measure a sample. If the spectrum of the opal glass is almost a straight line within a few percent of 100% reflectance and if the specimens have a variable, relatively low reflectance, then the standardization protocol is not likely to affect the outcome of an experiment, and data may reasonably be reported as being relative to opal glass, within X% of magnesium oxide. However, if the measurements themselves have an intrinsic importance (rather than being used to test an experimental effect), there is no option but to correct all the data relative to magnesium oxide. If done properly, this should enable another

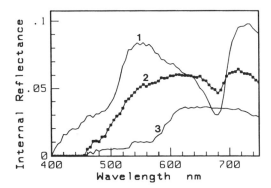

FIGURE 11.4
Internal reflectance spectra of green (1), yellow (2), and red (3) tomatoes measured around the stalk.

researcher using different apparatus to duplicate the measurement. Thus, standardization protocols should be taken very seriously for something like an ongoing quality control program for white textile or paper fibers but, for a one-time experiment encompassing biological variation, a more relaxed approach is forgivable.

Protection of the distal end of the light guide may be required. Fluid may be drawn inwards between the optical fibers by capillary action. If the cladding is in good shape, the effect should be minimal, but contamination of the apparatus can be a concern. Epoxy coatings over the optical fibers may solve this problem, but it is important to test the system by backing the light guide away from the standard. Ideally, the reflectance spectrum should stay completely flat as the probe is backed away, although at overall reduced reflectance. If the spectrum begins to develop a shape as the light guide is moved away from the standard, then the same spectral bias might be imposed on measurements. The effect can be seen quite strongly if the light guide is protected with a film such as Saran.

11.3.2   Examples

Figure 11.4 shows some internal reflectance spectra of red, yellow, and green tomatoes, made with the light guide around the stalk of the tomato where commercial measurements are made to evaluate ripeness. Another whole realm of fiberoptic applications is created once an optical fiber is combined with a reagent at its distal end to form an optrode (Seitz, 1984). Figure 11.5 shows the absorbance at 600 nm of phenaphthazine indicator paper localized on the distal end of an optical fiber in contact with buffer (Figure 11.5, 1) or in contact with indicator paper overlying skeletal muscle (Figure 11.5, 2), thus allowing the CAM to be used for monitoring pH. Reagents may be separated from tissue samples using a variety of techniques, such as a selectively permeable membrane or immobilization on sepharose or polyacrylamide (Peterson and Vurek, 1984).

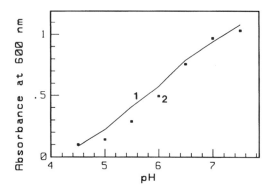

FIGURE 11.5
Absorbance at 600 nm of phenaphthazine pH indicator paper in buffer (1) and overlying skeletal muscle (2).

11.4   Spatial Scanning

In the layout shown in Figure 11.6, the two separate optical fibers connecting the CAM to the sample allow separation of the illuminating window from the receiving window. This might be done to create a long light path through tissues, which then would emphasize selective absorbance by chromophores in the liquid phase of the tissue. But instead of merely clamping the two windows at a fixed distance apart and with a fixed orientation, what happens if the illuminating window is held at a constant position under a slice of tissue, while the receiving window is scanned under computer control over the top of the tissue slice?

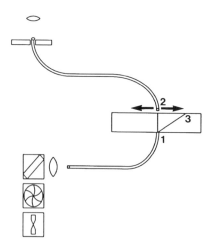

FIGURE 11.6
Scanning a tissue slice with the CAM. The optical fiber from the illuminator is fixed in the platform supporting the tissue slice (1), while the receiving optical fiber to the spectrophotometer is scanned across the top of the sample by a servomotor (2), thus detecting the extent of lateral scattering (3).

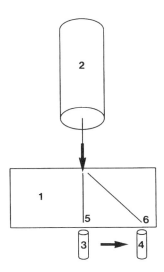

FIGURE 11.7
Laser scanning principle introduced by Birth et al. (1978). A slice of tissue (1) is illuminated from above by a laser (2), and the pattern of light transmitted is detected by a photodiode which scans from the optical axis (3) out to one side (4). The initial path length (5) increases as the angle through the tissue increases (6).

11.4.1   Monochromatic Scattering with a Laser

Equipped with optical fibers, a CAM may be used to investigate light scattering in complex tissues. Birth et al. (1978) showed that the spatial pattern of scattered laser light may be used to measure pH-related scattering in muscle. Striated muscle is composed of thick (myosin) and thin (actin) filaments held apart by negative electrostatic repulsion. When a muscle contracts, the filaments slide past each other. Most muscle fibers store glycogen as an energy source, providing a substrate for glycolysis. Under anaerobic conditions (once a muscle sample has been separated from its blood supply), lactate is formed by anaerobic glycolysis, causing the pH to decline. Thus, as the pH of muscle proteins declines towards their isoelectric point, electrostatic inter-filament repulsion is reduced, and filaments move closer together with an efflux of fluid. As living muscle becomes fatigued, therefore, it shows an increase in light scattering (Westerblad and Lännergren, 1990). The ultrastructural mechanisms involved are known from studies using X-ray diffraction and interference microscopy (Swatland et al., 1989), but scanning with the CAM allows us to examine these changes in bulk tissues.

The laser method developed by Birth et al. (1978) is based on the Kubelka-Munk analysis of ascending and descending fluxes of diffuse light in a translucent system (Judd and Wyszecki, 1975). The upper surface of a slice of muscle is illuminated with a red helium-neon red laser, and a photodiode scans across the lower surface (Figure 11.7). If scattering is minimal, the laser beam is transmitted directly with little lateral deflection to give a small, bright spot of light on the underside of the sample, whereas,

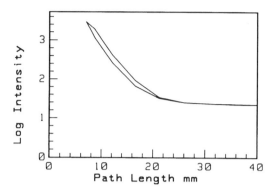

FIGURE 11.8
Laser scanning of a slice of skeletal muscle.

if scattering is high, the spot is larger and has a lower intensity because more light has been deflected sideways. This may be measured as follows:

$$\log M_T = A - Br$$

where M_T = radiant exitance on the lower surface, A = intercept or β_0 of linear regression, B = slope or β_1 of linear regression, and r = path length through the sample. Birth et al. (1978) showed that:

$$B = \log 2 \, (S + K)$$

where S = scatter coefficient (per cm) and K = absorption coefficient (per cm), as in the Kubelka-Munk analysis. Thus, B may be used to obtain the scatter coefficient combined with the absorption coefficient, but, unfortunately, these two coefficients are difficult to isolate separately. A typical result is shown in Figure 11.8.

11.4.2   Polychromatic Scattering with a CAM

For spectral studies, two changes to the CAM layout are useful. First, the tissue is illuminated from below, so that the tissue slice may be supported on a platform drilled to take the illuminating optical fiber from the CAM. Second, a white-light illuminator is used instead of monochromatic laser light. Using a PMT in place of a photodiode, it is now possible to examine spectral dispersion at various points across the spatial scan over the tissue. Thus, the primary scanner is the CAM spectrophotometer, and the secondary scanner is the spatial scanner over the top of the tissue slice.

This configuration yields a data matrix with wavelength on one axis and spatial position on the other, as shown in Figure 11.9, where two wavelength-position matrices are concatenated in the first dimension. The intensity of transmittance is shown by the degree of stippling. Sample 1 had a high pH and low scattering, while

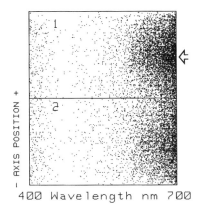

FIGURE 11.9

Gray maps of two wavelength-position matrices obtained with apparatus in Figure 11.7 for high-pH (1) and low-pH (2) skeletal muscle. The probability of a pixel being turned on is proportional to the light intensity. The two matrices had to be concatenated to get the same scaling for pixel probability vs. intensity. Thus, the most intense forward transmittance was in sample 1, at 700 nm, in the central axis (arrow). Violet light at 400 nm was scattered laterally to the axis to a similar extent in both samples. The least intense forward transmittance was around 500 nm in both samples, because of absorbance by myoglobin in the liquid phase of the tissue.

sample 2 had a low pH and high scattering. The wavelength-position matrix is the starting point for a detailed analysis of pH-related scattering, which begins to diverge from the subject of this book. But the main point is that scattering in tissue is strongly affected by wavelength, because scattering tends to be inversely proportional to the fourth power of wavelength. Thus, red light at 700 nm has a strong forward transmittance through tissue, while violet light at 400 nm tends to be scattered sideways. Using the spatial scattering principles expounded by Birth et al. (1978) at different wavelengths provides a more sensitive indicator of scattering than does a single wavelength, as well as providing data for multivariate analysis.

11.5   Goniospectrophotometry

A goniometer is an instrument for measuring geometrical angles, and a goniospectro-photometer is a device for measuring spectra at different angles. Figure 11.10 shows the principle of fiberoptic goniospectrophotometry for measuring light scattering in bulk tissues. The optical fibers are fitted tightly into hypodermic needles inserted in a radial pattern through a jig into the tissue, like the spokes of a wheel. One optical fiber illuminates the tissue at the center of a circle, around which the other optical fibers collect light equidistantly from the tip of the illuminating optical fiber. The receiving optical fibers terminate on the scanning stage of the CAM, where their coordinates have been mapped for repeated measurement. Thus, the primary scanner is the CAM spec-trophotometer while the secondary scanner is the scanning stage.

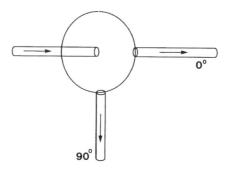

FIGURE 11.10

Fiberoptic goniospectrophotometry. The illuminating optical fiber connects to the CAM illuminator, and receiving optical fibers connect to the scanning stage of the CAM (only two are shown, at 0° and 90°).

11.5.1   Hardware

The angle of the optical fiber collecting light directly in line with the illuminating fiber is defined as 0°. Thus, an optical fiber perpendicular to both the illuminating fiber and the 0° fiber collects light at 90° to the light passing straight forward through the tissue. The light passing straight forward through the tissue (out of the illuminating optical fiber and straight into the 0° fiber) is the strongest relative to other angles, particularly at 700 nm (Figure 11.11, 0°). Light scattered sideways is weaker than the straight forward light, particularly at 700 nm (Figure 11.11, 90°). However, at 400 nm, approximately the same intensity of light enters all the optical fibers, from 0° to 90° regardless of angle, because low wavelengths tend to be uniformly scattered through the tissue, whereas high wavelengths are scattered less and have a higher forward transmittance.

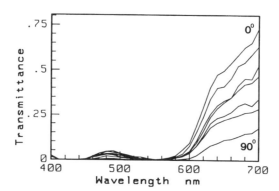

FIGURE 11.11

Fiberoptic goniospectrophotometry of skeletal muscle, showing spectra collected at increasing angles from 0° to 90°.

11.5.2   Standardization

If the light path through the tissue is much more than a few millimeters or if the tissue is heavily pigmented, the light intensity reaching the PMT may be quite low. As may be seen from Figure 11.11, experimental error may become quite noticeable. This can be reduced by averaging the results from replicate needles at comparable angles at each side of the main optical axis through the tissue; however, this does not improve the reliability of measurements made from the single needle at 0°. Another problem is deciding on a standard. If the 0° optical fiber is used to standardize the system, there may be a complex interaction with the absorbance spectrum, similar to that described in Chapter 8 for the polarization of reflectance measured with a tilting stage. The data in Figure 11.11 were standardized with the needles in place around the jig, but with no sample in position. This results in a fiber → air → fiber linkage which may be quite different from the fiber → tissue → fiber linkage used for measurements. However, if the illuminating optical fiber is standardized in a repeatable manner against opal glass, then most of the dynamic range of the PMT is wasted because the standardization intensity is orders of magnitude greater than the light that penetrates through tissues. Thus, standardization is a somewhat of a problem.

11.6   Remote Fluorometry

The normal configuration of the CAM for fluorometry is to use an epi-illuminator for excitation. Thus, the substage condenser may be removed and an optical fiber used as the target for the objective. The excitation then passes along the optical fiber to its distal window in contact with the tissue. Any fluorescence at the distal window travels back along the optical fiber, through the dichroic mirror, and onwards to the spectrophotometer. Thus, remote fluorometry is even simpler than remote spectrophotometry. There are many different applications, such as perfusion fluorometry (Shackleford and Yielding, 1987) and in monitoring the state of living tissues during experiments (Udenfriend, 1969). For example, surface fluorometry may be used for monitoring the glycolytic status during heart failure (Auffermann et al., 1990). Figure 11.12 shows a less dynamic example, the remote measurement of lignification in sliced broccoli stems.

Remote fluorometry can be combined with remote spatial scanning to produce another type of wavelength-position matrix. Figure 11.13 shows a stack of fluorescence emission spectra (with excitation at 365 nm) made at 2-mm intervals across a slice of skeletal muscle. The centrally located emission peak corresponds to the intramuscular tendon of the muscle. Thus, the CAM spectrophotometer was the primary scanner, and the servomotor for remote spatial scanning was the secondary scanner. But, what happens if the single optical fiber used for fluorometry is mounted in a hypodermic needle and pushed into the tissue instead of being scanned over its surface?

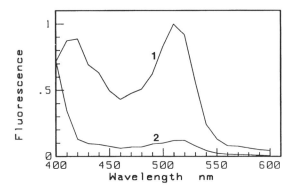

FIGURE 11.12
Fluorescence emission spectra of lignification in hard (1) and soft (2) broccoli stems.

11.6.1 Needle Penetration Fluorometry

Back in Chapter 1, when introducing the concept of primary and secondary CAM
scanners and the need for flexible programming, it was anticipated that spectra might
not always be regular or form a logical ascending sequence. When light-guide probes
or hypodermic needles with a single optical fiber are pushed into soft tissue, the
optical window depth may be monitored from a sensor geared to a flat plate remaining at the surface. Depth could be incremented in regular steps with a stepper motor,
or the photometer of a hand-held probe could be triggered from an optical-shaft
encoder used as the depth sensor. But the most simple method is to alternate
photometer measurements with depth measurements from a continuously variable
potentiometer, reversing the order of measurement as the probe is withdrawn to avoid
hysteresis, in which case the number of measurements and the scanning increment
(depth) may be irregular. Dynamic analysis of probe movements reveals that, unknown to the operator, there usually is an interaction with the human cortical and
cerebellar control of constant velocity movements by the arm. The tip of the probe

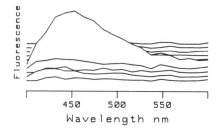

FIGURE 11.13
Fluorescence emission spectra of intramuscular collagen encountered with an optical fiber stepping at 2-mm
intervals across the surface of a slice of skeletal muscle.

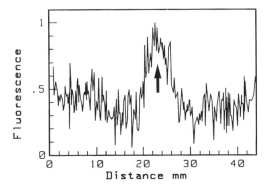

FIGURE 11.14
Signal from a fiberoptic hypodermic-needle fluorescence probe pushed into a seal liver with a subsurface fibrotic lesion (arrow). The distance measured is the depth of the optical window in the tissue.

may pause or even recoil backwards as it encounters zones of mechanical resistance in the tissue. Furthermore, measurements may be taken as the probe is pushed into the tissue, to be compared with measurements taken as the probe is withdrawn from the tissue. The difference between the two signals gives an indication of the degree of tissue deformation during initial penetration.

Figure 11.14 shows an example of this method used in the detection of fibrotic lesions in seal livers (collected to explore interactions between marine pollution and wildlife pathology). The small peaks were generated by the normal connective tissue matrix present in both normal and diseased livers, but broad-band (440- to 600-nm) fluorescence emission rose as the needle transected a lesion. Thus, there was no significant difference between normal and fibrotic livers for the numbers or widths of fluorescence peaks, but fibrotic livers had a higher mean fluorescence (raw PMT output, 1217 ± 1.58 vs. 1222 ± 1.58, $P < 0.01$). Relative to programming for a CAM, this example shows how step-and-measure types of analysis eventually may evolve into signal analysis. Thus, software for vectors of variable length with variable degrees of freedom is far more flexible than software based on fixed matrices.

11.6.2   Vignette Window Effect

When a fiberoptic needle probe is pushed into a tissue, the information is acquired sequentially through a small optical window which vignettes anatomical structures larger than the window as they pass. This effect is complicated still further if there are differences in the luminous intensity of structures passing the window. To explain the point, imagine the sun shining through the window of a room onto a large photometer. The window has a roller blind of opaque material that can be drawn down to block the light. By experiment, a photometer reading could be found at half light-intensity, corresponding to the position of the blind half-way down the window, which then could be used to identify the half-way position. Thus, the PC could be programmed to distinguish between blind-up and blind-down positions or, in penetrating a tissue, to count the number of luminous structures passing the probe window.

Returning to the analogy of the window blind, if a translucent blind is substituted for the opaque blind, the photometer will be biased for finding the position of the blind. The translucent blind will be more than half-way down the window before the photometer threshold (half light intensity) is reached. Similarly, a peak-counting algorithm in a fiberoptic probe will fail if it cannot cope with differences in the luminous intensity of the structures to be counted and measured in width (such as the peaks shown in Figure 11.14). Thus, counting and measuring the widths of peaks in a biological tissue may not be as simple as they might first appear. Not only can there be differences in luminosity of the item to be detected, but the background luminosity may also vary with pH or lipid content.

A fiberoptic window in a probe may be made as small as possible, relative to the anatomical structure to be detected. And the lower size limit of optical fibers is certainly very low. But this causes other problems. First, the small window will be overly sensitive to all small tissue inclusions, not just those to be measured. Second, the amount of light reaching the photometer will be reduced, thus requiring greater amplification which, in turn, increases the noise of the signal. As the noise level rises, it becomes progressively more difficult to recognize the anatomical structure. To overcome this problem, the intensity of the light source might be increased, but the photometer now will react to an enlarged cone of diffuse light outside the window, so that the operational window size is increased again, looping back to the start of the problem.

Plastic optical fibers that can sustain a small radius of curvature are limited in the wavelengths they transmit, while relatively rigid quartz fibers with high transmittance at low wavelengths can only be used at a large radius of curvature. Thus, a side-mounted window in a probe is likely to be elliptical, although it may be modeled as a circle, as shown on the right of Figure 11.15. This geometry also applies to a group of optical fibers in a light guide collectively forming a window with an elliptical or circular outline. For further simplification, it is assumed that the single optical window acts both to illuminate and to measure, as in a fluorometry probe.

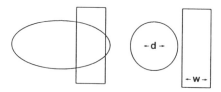

FIGURE 11.15

Geometrical models for signals created at an optical window. Probe windows often form an ellipse (left), but these can be modeled as a circle (diameter, d) passing a rectilinear bar (width, w) representing anatomical structures such as connective tissue layers and blood vessel walls.

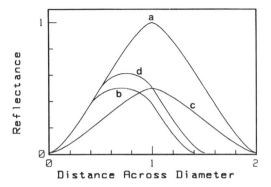

FIGURE 11.16

Signals predicted for rectilinear bars of varying width (w) and reflectance (r) passing across a circular window, diameter = 1. The position on the window diameter reached by the leading edge of the bar is shown on the x-axis. A bar with w = 1 and r = 1 is shown by line a, with w = .4 and r = 1 shown by line b, w = 1 and r = .5 shown by line c, and w = .5 and r = 1 shown by line d.

If the window is circular with a known diameter (d), and the complexity of anatomical shapes passing the window is modeled by a rectilinear bar with a known width (w), then a photometer measuring the reflectance (r) or fluorescence of the bar through the window will produce a signal similar to that shown by line a in Figure 11.16 for d = 1, w = 1, and r = 1. The signal rises as the leading edge of the bar enters the window, reaches a maximum when the bar fills the window, and declines as the trailing edge of the bar leaves the window.

Comparison of lines b and c in Figure 11.16 shows that, by itself, the amplitude of reflectance cannot be used to distinguish between bars that differ in both width and reflectance. Furthermore, a bar with w = .5 and r = 1 attains a reflectance > .5 because it fills more than half the window area when the center of the bar is at the center of the window (line d in Figure 11.16). Thus, with real signals from complex anatomical structures in the tissue, both the amplitude of the signal and a measure of its peak width are required to resolve the anatomical feature passing the window.

A variety of factors affect the shapes of real signals. Apart from the complex shapes of many anatomical structures such as blood vessels, nerves, and layers of connective tissue, smearing may occur across the optical window as the tissue is deformed or disrupted by the probe. The response time of the photometer also may produce a similar distortion of signal shape as a function of probe velocity vs. photometer response time. Bearing in mind that the probe moves in a straight line through the tissue, the most important dimensions of windows and measured objects are their effective diameters parallel to the probe direction.

Predictions may be tested experimentally by measuring the reflectance of white bars against a black background, moving a test chart past the optical window of a probe. Measured values may be close to predicted values, as shown in Figure 11.17. But, testing also may reveal errors or further complexities of the window, as in Figure 11.18, where the difference between predicted and actual signals was caused by the

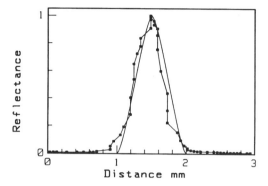

FIGURE 11.17
Experimental detection of a white test bar (r = 1, w = .5 mm) against a black background by a single optical fiber (d = .5 mm) in a light guide window, with illumination of the test bar from other fibers in the light guide. Predicted and measured signals are shown by the simple line and the line with squares, respectively.

non-random arrangement of optical fibers in a light guide. Half the fibers were used to illuminate the bar, while the remainder were used to monitor its position.

Figure 11.19 illustrates a window problem originating from the curvature of an optical fiber to form a side window on a probe shaft. Placing the elliptical window close to the curvature of the optical fiber within the probe shaft increases the light intensity at the outer radius of the curve, so that light radiates from the optical window asymmetrically. Thus, in Figure 11.20, the light radiated forward in advance of the window and detected the bar before it actually reached the window. When testing a probe window, therefore, it is useful to measure the light intensity radiating from the window by moving a photometer in an arc around the window. In Figure

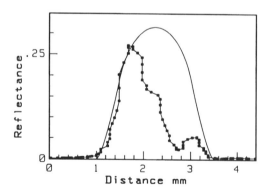

FIGURE 11.18
Detection of a test bar (r = 1, w = .5 mm) by half the fibers of a light guide (d = 2 mm), with illumination by the remaining fibers. Predicted and measured signals are shown by the simple line and the line with squares, respectively.

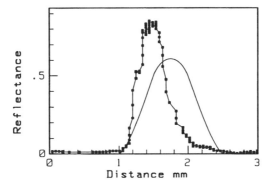

FIGURE 11.19
Detection of a test bar (r = 1, w = .5 mm) by a single, large-diameter optical fiber (d = 1 mm), with light passing in both directions in the same fiber. Predicted and measured signals are shown by the simple line and the line with squares, respectively.

11.20, for example, the asymmetry of the optical window that produced the effect shown in Figure 11.19 may be seen.

One of the important consequences of the vignette effect in understanding signals from tissue probes is the relationship between the size of an anatomical structure and the base width of the peak it creates in the probe signal. If an anatomical structure has a narrow width relative to the diameter of the window, the base width of its peak is a little larger than the diameter of the window. For example, measured in relative units of window radii, with a window diameter of $d = 2$ and a bar width of $w = 0.2$, then the base width is $d + w = 2.2$, as shown by the first line in Figure 11.21, and the peak intensity of the signal never reaches the maximum value that

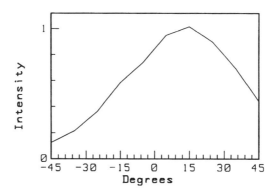

FIGURE 11.20
Goniophotometry of light radiating from an optical fiber on a probe. The 0° position shows light radiating perpendicular to the shaft of the probe; negative degrees are towards the handle of the probe, and positive degrees are towards the tip of the probe.

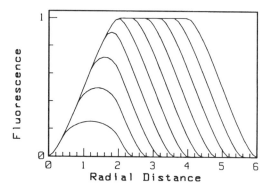

FIGURE 11.21

Test bars of different width (0.2 to 2 window radii in width) but equal fluorescence (fluorescence = 1) pass
an optical window to create peaks in the probe signal.

would occur if the structure completely filled the window. When the width of the
anatomical structure equals the diameter of the window (second line in Figure 11.21),
the base width of the signal is twice the diameter of the window but still is given by
d + w = 2 + 2 = 4. The relationship also holds for structures that are large relative
to the window, as for the case of a bar with a width of 4 radii (d + w = 2 + 4 = 6;
third line in Figure 11.21).

11.6.3   Spectral Window Effect

Optical fibers have many applications in probing tissues, as clearly envisaged by the
pioneers in the field (Kapany, 1967). When optical fibers are interfaced directly with
tissues, however, it is difficult to predict the effects of wavelength on the optical
properties of the window and in the tissues beyond. The refractive index of the fluid
phase of a biological sample may affect the Fresnel reflectance losses at the optical
window, which are comparable in principle to those that occur in the coupling of
optical fibers used for communication (Hewlett-Packard, 1982). When passing from
one medium (x) to another (y), with refractive indices n_x and n_y, respectively, then:

$$\text{Fresnel loss (dB)} = 10 \log ((2 + (n_x/n_y) + (n_y/n_x))/4)$$

Since both the refractive index and transmittance of optical fibers are related to
wavelength, it is difficult to predict how white light will be transmitted into the liquid
phase of tissues varying in refractive index.

   When optical fibers are used for communication, they are cut perpendicularly
to the long axis of the fiber, unless a special type of coupling is in use, but in a
penetration probe the optical window may be an ellipse immediately behind the
cutting tip of a hypodermic needle or an ellipse on the side of a light guide probe.
This adds another dimension to the problem of understanding the optical window.

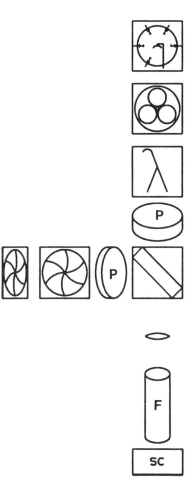

FIGURE 11.22
CAM layout for examining the internal reflectance at the distal window of an optical fiber. Crossed polarizers (P) remove reflectance at the top or distal window of the optical fiber (F), so that the CAM is coupled with the distal window of the optical fiber, which is inserted into a sample chamber (SC) containing fluids of different refractive index.

11.6.4   Internal Reflections

A CAM layout such as that shown in Figure 11.22 allows internal reflectance of the distal optical window to be measured. The logic of the system is that, because it is difficult to get into the tissue to measure the light coming out of the optical window, then the opposite information may be obtained within the apparatus by measuring the internal reflectance at the distal window.

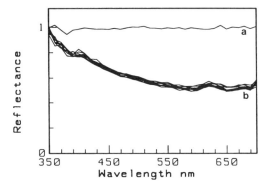

FIGURE 11.23
Internal reflectance from the distal window (90°) of an optical fiber standardized in air (a) and then inserted into fluid (b) ranging from water (n = 1.333) to approximately 63.5% sucrose (n = 1.450).

If the distal window of an optical fiber is cut at 90° to the longitudinal axis, standardized to give a reflectance of 1 in air (n = 1), and then placed into distilled water, there is a decrease in internally reflected light (from a to b in Figure 11.23). The large change in internal reflectance from air to water shows why standardization of a fiberoptic system from a white standard in air may be misleading when measurements are to be made with the optical fiber pushed into wet tissue. But the differences between water (n = 1.333) and 63.5% sucrose (n = 1.450) are far less, so that small differences in the refractive index of tissue fluids were not important for the plastic optical fiber on which these measurements were made. However, the shape and magnitude of this effect depend on the type of optical fiber and should be checked for a window likely to be exposed to fluids differing in refractive index.

11.6.5   Hypodermic Needle Angle

The angle at which the distal window of an optical fiber is cut (relative to the longitudinal axis of the optical fiber) also may have a strong effect on internal reflectance. If an optical fiber cut at 90° is used to standardize the spectrophotometer, shaving wedges off the perpendicular end face first reduces the internal reflectance, then increases it. This has some important consequences when an optical fiber is fitted into a typical 15° hypodermic needle. The most effective angle for the cutting tip may or may not be an efficient angle for the optical fiber, depending on the refractive index properties of the fiber.

11.6.6   Widespread Occurrence of Aperture Effects

In conclusion, it may be of interest to note that many of the problems discussed here with reference to fiberoptic probes are generic in nature. They often appear when structural measurements are made through an aperture, if either the structure or the aperture is moving. For example, back in the days before inexpensive frame-grabber boards were available for VIA, methods were developed to count anatomical structures such as boundaries between muscle fibers, using a scanning stage and a hard-wired counting circuit (Swatland, 1979). The same set of problems was encountered. Thus, the underlying geometry shown in Figure 11.15 also applies to the relationship of slit width vs. photometric aperture in a monochromator (Chapter 5).

11.7   Coherent Fiberoptic Light Guide for Remote Imaging

In this simple but useful method, the layout of the CAM follows that shown in Figure 11.3, except that the receiving optical fibers are coherent. There is a lens to focus the image of the specimen onto the coherent light guide. When the other end of the coherent light guide on the LM stage is examined with a low-power objective, the coherent optical fibers appear as a precise matrix. Spread over the whole matrix can be seen the image of the specimen. Having selected part of the specimen to measure, the appropriate optical fiber corresponding to that part of the image is moved to the optical axis. The objective magnification is increased and, using an appropriate photometric aperture, light from this one optical fiber is directed into the optical axis of the CAM. The method is ideal for measuring small flecks of color (0.5 to several millimeters) in a heterogeneous system that cannot reasonably be placed on the LM stage. The observer can check that the correct area is being measured relative to the whole of the specimen.

11.8   Polarized Light

When used in a normal manner, optical fibers rotate light as it is reflected around numerous bends. Thus, they cannot directly be used to extend the working range of the polarizing CAM. Polarization-preserving optical fibers are available but, at present, they are difficult to use on a CAM (they work best with a laser); however, miniature polarizers can be fitted to the ends of the distal ends of optical fibers where they interface with the specimen. If the light is passing through the specimen from one side to the other, with a fixed polarizer before the specimen and a rotatable analyzer after the specimen, the maximum transmittance occurs when the analyzer is rotated parallel to the polarizer, and the minimum is when the analyzer is perpendicular to the polarizer. But if the polarizer is at 0°, when the analyzer is at 90°, then a birefringent structure at 45° between the polarizer and analyzer rotates light so that now it can pass through the analyzer (Chapter 7). These effects may be measured in tissue slices using a CAM layout such as that in Figure 11.6 if scattering is minimized by using relatively high wavelengths in the NIR region of the spectrum.

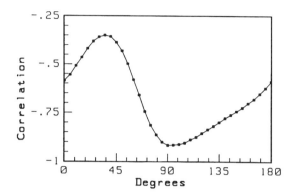

FIGURE 11.24

Angular distribution of the coefficient of correlation of transmittance with sarcomere length in 1-mm thick slices of skeletal muscle.

Light microscope polarizers intended for visible light cannot automatically be assumed to work for NIR, so this must be checked by calculating extinction coefficients at different wavelengths:

$$k = \log_{10}(T_0/T_{90})$$

where T_0 is with the analyzer parallel to the polarizer, and T_{90} is with the analyzer perpendicular to the polarizer. For example, polarizers on a typical polarizing LM might have values for k at increasing wavelengths of 1.21 at 500 nm, 1.52 at 600 nm, 1.69 at 700 nm, 1.65 at 750 nm, 1.56 at 800 nm, 0.43 at 850 nm, and 0.05 at 900 nm. Thus, at 800 nm, this polarizer-analyzer combination could be used as effectively as with visible light, but not above 800 nm.

11.8.1 Transmittance

Thick slices of tissue measured with fiberoptic light guides exhibit many of the properties detectable in histological sections of the same tissue. Working with skeletal muscle, samples were fixed in 20% w/v isopropyl alcohol, 4% w/v paraformaldehyde, and 2% w/v trichloroacetic acid for 3 days, before being sliced at a thickness of 1 mm. Sarcomere length was measured by diffraction with a helium-neon laser (Rome, 1967) and checked against measurements made by differential interference contrast (objective ×100, NA 1.25). Only a small difference (0.02 μm) was detected. Transmittance through the sample at different analyzer angles was correlated with sarcomere length, as shown in Figure 11.24. Thus, it is possible to detect birefringence with relatively thick tissue slices, just as it is with a histological section on the stage of CAM. In Figure 11.24, samples with short sarcomeres had higher birefringence, giving higher transmittance through crossed polarizers at 90° and a negative correlation of sarcomere length with transmittance.

11.8.2   Back-Scattering

It is difficult to cut thick sections of fresh tissue at an accurate thickness. So, instead of passing light through the sections, it is easier to obtain similar but opposite results by detecting back-scattering from intact tissue. Crossed polarizers were mounted directly on a split-diameter bifurcated light guide (Figure 11.2, 4) in contact with pre-rigor muscle. While sarcomere length was still extensible, muscles were clamped at their natural rest length, then stretched to 164% rest length. Without the polarizers, stretching produced no detectable effect on back-scattering of NIR. But with the crossed polarizers in position, stretching increased (P < 0.001) detectable levels of back-scattered NIR. Thus, by using back-scattering rather than forward transmittance, fiberoptics may be used to take some of the operations of the polarizing CAM directly to the surface of a tissue (Swatland, 1996b).

12

Programming the Sample Environment

12.1   Introduction

If a sample can be trapped or secured in a small transparent chamber, it can be examined and measured directly on the LM stage (Diller, 1982; McGrath, 1985; Finch and Stier, 1988; Braga, 1990). Thus, sometimes it is possible to observe cells and small organisms without killing them by fixation, sectioning, staining, and embedding. Direct observation is particularly useful for easily immobilized specimens, such as nerve, muscle, connective tissue, or textile fibers. However, direct observation may entail a loss of resolution or magnification because of the depth of the sample chamber, as discussed in Chapter 1. Living cells also may be examined in a radial flow chamber, injecting pulses of biologically active materials to examine cellular responses (Ruzicka et al., 1996), or monitoring moving cells in a fluid gradient chamber (Ebrahimzadeh et al., 1996).

The goal of this final chapter is to explore ways to manipulate the sample environment. As far as CAM software is concerned, a programmable sample environment is just another dimension of scanning through which to step-and-measure.

12.2   Regulating pH

Many of the optical properties of tissues, such as fluorescence absorbance spectra (Krammer and Uberriegler, 1996) and birefringence (Chapter 7), depend upon pH. Regulating pH is a routine method in many manufacturing industries, but there are some special considerations when it is done on the LM stage (Swatland, 1994). At first sight, there may appear to be many different ways to control the fluid environment around a microscopic specimen. Why not have two large reservoirs containing

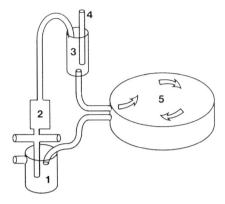

FIGURE 12.1
Overall plan of pH controller showing sump (1), pump (2), header tank (3), pH electrode (4), and sample chamber on the CAM stage (5).

the components of a buffer mounted above the microscope, with computer-operated solenoid valves controlling the mixture ratio for a gravity feed through the sample chamber? This arrangement may be made to work, but it is difficult to stabilize and is prone to flooding. To preserve numerical aperture as far as possible, only a thin cover slip can be used for the top of the sample chamber, and this may become a sensitive barometer to changes in hydrostatic pressure caused by valve operation. When the valves are opened the pressure in the chamber may cause the cover slip to bulge outwards or leak but, when the valves are shut, the cover slip may bulge inwards or air bubbles may be sucked in. Immersion objectives might be used with a sample chamber mounted directly on the objective, but then changing the magnification or positioning the sample would become a major operation. Thus, the solution to one problem becomes the source of others.

Designing a system for programmed pH changes while viewing a sample may be relatively simple, but the requirements for optical measurements are more stringent. One possible route through this labyrinth of design constraints is discussed here, reaching a compromise of complexity with performance and enabling optical measurements to be made directly on a stabilized sample.

12.2.1   Hardware

A primary design rule is to avoid the use of large gravity-feed fluid reservoirs above the level of any of the major electrical components, particularly the high voltage supply to the PMT, for obvious reasons. An overall plan is given in Figure 12.1. Fluid from a sump (1) is moved by a pump (2) to a header tank (3) containing a pH electrode (4) and then passes through the sample chamber (5) on the microscope stage. The header tank holds only a few milliliters of fluid at any time and is located 10 cm above the level of the LM stage. The flow rate through the chamber (0.5 ml/min) is determined by the tubing diameter after the header tank, the depth of fluid in the header tank, and the height of the header tank above the stage, with the latter being most easily adjustable.

Changes in pH may be created by opening solenoid valves in supply lines to the pump intake, with the advantage that pump stroke volume is the major determinant of fluid volume admitted per second of valve open time (0.16 sec/ml). The relative heights of reservoir valves above the sump fluid level (approximately 15 cm) and tube diameters (ID = 4 mm) are balanced so that hydrostatic pressure in the reservoir tanks has a minimal effect on the time:volume relationship of valves or back-flow into the sump between pump cycles. This may be accepted as a trade-off against the high cost and complexity of injector pumps, allowing surplus automotive valves to be used (such as Edmund Scientific N42533). Excess fluid is discarded through the overflow from the sump, at which level the whole system has a volume of approximately 100 ml; however, only 95 ml can be pumped out.

Fluid pH is monitored from a gel-filled combination electrode in the header tank (Figure 12.1, 4). Ease of removal for standardization against buffers makes this a better location than in the relatively inaccessible sump. It also provides a more immediate measure of the fluid passing through the sample chamber. A stainless steel mesh (square weave, 0.38-mm openings, #40 mesh) is located on the floor of the header tank to remove bubbles. Fluid volume in the system is maintained so that the fluid level does not drop below this mesh once the system is in operation.

Additional tubes and valves may be required. A large diameter (ID = 7 mm) tube from the header tank to the sump can reduce the pressure fluctuations in the sample chamber, as well as reducing cycling times and increasing fluid mixing (Figure 12.2, 1). With its upper opening above the level of the de-bubbling mesh, it acts as an overflow for the header tank. A continuous flow cell refractometer (such as the Zeiss model A Abbe refractometer) may be placed in series with this tube, with the flow cell as a "U" bend trap.

Complete isolation of the sample chamber is required when optical measurements are made, using two solenoid-operated precision valves to prevent moving fluid from causing sample movements. Isolation of the chamber also ensures that the cover slip remains as flat as possible. The upper valve is a simple shut off (Figure 12.2, 2). The lower valve (Figure 12.2, 3) is a two-way valve that shuts off the sample chamber but opens a small-diameter flushing tube from the header tank (Figure 12.2, 4). At the edges of the buffering range, it is important to minimize the unflushed volume in the apparatus. And, with this valve arrangement, optical measurements can be made in the isolated sample chamber by the main controller while a secondary controller adjusts the pH of fluid outside the chamber to its next programmed value.

An exit valve may be added just after the pump to reduce wastage and facilitate calculations. This is used to dump fluid out of the system, prior to an equal volume of fluid being introduced into the system through acid or base valves. The header tank and sump both have high-level detectors generating non-maskable interrupts to their immediate controller.

A relatively simple sample chamber may be constructed from a plastic ring (Figure 12.1, 5; ID = 18 mm, height = 4 mm) with two tubes for the inlet and outlet

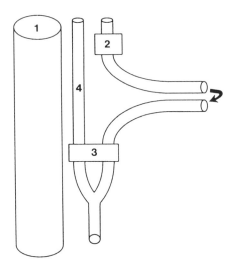

FIGURE 12.2
Extra valves and tubing for the pH controller. A large diameter tube overflows from the header tank to
the sump (1); a shut-off solenoid valve (2) isolates the sample chamber (arrow); a two-way valve (3) either
drains the sample chamber or drains the header tank via a small-diameter tube (4) when the sample
chamber is closed.

of fluid (ID = 0.5 mm). Both tubes should be mounted next to each other to create
a circular flow of fluid around the chamber (which is less likely to displace the
sample than a current across the chamber). An air bubble is likely to be created when
the sample is first sealed into the chamber, using rubber cement between a normal
glass microscope slide and cover slip (0.17-mm thickness). But, if the isolating
valves (Figure 12.2, 2 and 3) are opened (preferably by local switches overriding
program control), the slide can be tilted with inlet and outlet tubes uppermost and the
bubble sucked out. Although the chamber depth is 4 mm, the sample is secured to a
small metal frame just under the cover slip. Long-distance objectives (such as Zeiss
series LD-Epiplan SM objectives) may be used for parts of the sample not immediately under the cover slip. Although designed for a protective cover glass with a
thickness of 1.5 mm, Zeiss LD-Epiplan objectives with a 0.17-mm cover slip plus a
depth of water do not suffer a noticeable loss of resolution, bearing in mind that
resolution is already low relative to a permanently mounted thin section of tissue.

12.2.2   Software

Sørensen's phosphate buffer is useful for biological samples and serves as a programming example, as shown in Figure 12.3, where programming pH is reduced to
balancing the volumetric ratios of equimolar solutions of dibasic sodium phosphate
and monobasic potassium phosphate (Cooper, 1977):

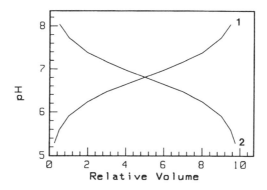

FIGURE 12.3
Volumetric relationships of stock solutions of Na_2HPO_4 (1) and KH_2PO_4 (2) in Sørensen's phosphate buffer.

$$pH = pK\alpha_2 + \log \frac{[HPO_4^{2-}]}{[H_2PO_4^-]}$$

Problems caused by unflushed volume at the edge of the buffering range may be seen where small volumes of the minor component cause large changes in pH. The buffer is being used primarily to change the pH around a relatively nonreactive sample, not actually to buffer the pH of the sample. Thus, it is permissible to use the edges of the buffering range.

Figure 12.4 shows an example of a programmed pH change, from pH 6 to 7. The status of the acid (A), base (B), and exit (E) valves is shown in the upper part of the figure. The volume of existing fluid to be replaced by base was found from the relationship shown in Figure 12.3. The cycling and mixing time of the system (22 sec) is shown by the arrows in Figure 12.4.

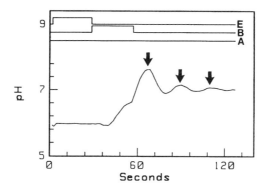

FIGURE 12.4
A one-shot programmed change from pH 6 to 7, showing the status of acid (A), base (B), and exit (E) valves.

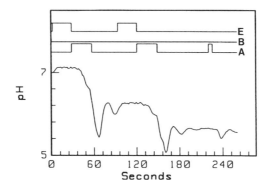

FIGURE 12.5

An iterative procedure change from pH 7 to 5.5, showing the status of acid (A), base (B), and exit (E) valves.

Direct calculation and one-shot solutions are not very reliable in practice where the volumes of moving and replaced fluid are known only approximately. A fairly long wait between changing the pH and measuring the sample is required, to be sure that equilibration has occurred within the sample. Thus, an iterative procedure is more robust, using a series of low estimates to reach the required pH, as shown in Figure 12.5, for a programmed change from pH 7 to 5.5. As seen from the valve status in the upper part of Figure 12.5, fluid was only dumped from the exit valve for the first two replacements by acid salt solution, so that the effect of the third addition was weakened to give some fine-tuning.

Although sacrificing speed of change, iterative procedures have many advantages over one-shot solutions. It is not necessary to know the exact chemical nature or dilutions of the buffer stock solutions, which facilitates preparation of large volumes of stock fluids. The first addition of acidic or basic components may be used to interrogate the system, finding the magnitude of the response for a known change and allowing the development of a simple expert system. But, just as in setting a PMT (Chapter 4), the target for the iterative procedure must be defined as an acceptance window rather than an exact value; otherwise, it may never be attained and the system may oscillate.

The array passed into the subprogram for changing the pH was as follows:

{1}   Test flag for printing operational status
{2}   Code for the identity of the buffer
{3}   Code for the instruction
     (1)   Programmed change to a target pH
     (2)   Read pH only
     (3)   Decrease pH by an arbitrary amount, wait, and read
     (4)   Increase pH by an arbitrary amount, wait, and read
     (5)   Dump fluid
     (6)   Display an on-screen window of pH and valve opening times, as in Figures 12.4 and 12.5

{4}   Flag to show that the system has been initialized (apparatus turned on and tested, fluid reservoirs filled, pH electrode standardized, etc.)

{5}   pH target for a programmed change

{6}   Width of the acceptance window, ± target

{7}   Duration for valve opening for arbitrary change pH and read options used in a simple step-and-measure algorithm

{8}   Most recently read pH value

{9}   Wait time for equilibration (number of cycles to read pH and wait 1 sec)

{10} Flag to show the original direction of change to reach a pH target (separate from the directional flag discussed later in planning the measuring loop): At the start of each programmed pH change, this conveys the same information as the current pH {8} compared with the target {5} ± window {6}, but it changes to reverse logic if the target window is overshot, which might occur if a narrow window is specified. This flag then may be used to lengthen the equilibration time, effectively damping oscillations before making an appropriate recovery.

{11} Actual volume of the system, 100 ml, used for initial calculations, before acquiring operational experience

{12} Apparent volume of the system (updated from operational experience) when increasing the pH

{13} Apparent volume of the system (updated from operational experience) when decreasing the pH: Ascending and descending apparent volumes may differ, {12} vs. {13}, because they describe the operation of different valves and stock solutions. Both {12} and {13} tend to increase relative to {11} as operating experience is gained. This may be caused by a number of factors such as stiff valves admitting less fluid than intended or the accidental loss of newly added rather than fully mixed fluid through the exit valve or overflow.

An algorithm to reach a programmed pH target (± window) and to make an optical measurement (absorbance, optical path difference, size, fluorescence, etc.) is as follows:

[1]   Find the current pH and, using the model of the buffer system (as in Figure 12.3) together with a correction from previous operational experience, if available, make a low estimate of the volume of fluid to be replaced by acid or base to reach the target pH.

[2]   Convert volumes to valve open times, replace fluid, and wait for equilibration.

[3]   Find the current pH and, if it is outside the target window, go to [1].

[4]   Wait again.

[5]   Find the current pH and, if it is inside the target window, go to [8].

[6]   Using the difference between the current pH and the target, corrected by previous operational experience if available, make a low estimate of the volume to be added to reach the target pH.

[7]   Convert volume to valve open time, add fluid, wait for equilibration, and go to [5].

[8]   Wait for the time required for internal equilibration of the sample (which usually is too small in volume to produce a major change in system pH).

[9]   Isolate the sample chamber (valves 2 and 3 in Figure 12.2), trapping fluid of a known and stable pH inside.

[10]   Measure the sample.

In Figure 12.5, for example, the first two openings of the acid valve were for fluid replacement and were called from step [2], while the third opening was for fluid addition and was called from step [7]. Using the small-diameter tube that bypasses the sample chamber when it was isolated (Figure 12.2), often it was possible to do two things simultaneously:

(1)   Internal equilibration and optical measurement of the sample at a previous pH

(2)   Change the remainder of the system to the next pH

Near the middle of the buffering range (Figure 12.3), reconnection of the sample chamber to the system causes only a small change in pH, which usually is within the acceptance window.

In operation, when the pH was changed from 5.5 to 7 there was a small increase in the refractive index of the phosphate buffer (from n = 1.3330 to n = 1.3342). This could be a problem in critical applications, which can be solved by adjusting the refractive index of the medium, as described later.

12.2.3   Example

Striated skeletal muscle fibers are essentially cylindrical in shape when released by dissection. The major striations are anisotropic (A) and isotropic (I) bands, which differ in refractive index and protein composition (Huxley and Niedergerke, 1954; Huxley and Hanson, 1957). The striations are originally perpendicular to fiber length but readily become skewed by dissection so that several striations may be crossed diagonally while scanning across the width of a fiber mounted in the sample chamber.

Figure 12.6 shows a muscle fiber at two pH values: at pH 5, which is very low, and pH 6.5, which is quite high, relative to normal postmortem pH reached after glycolysis. The fiber was scanned using a mechanical stage (Zeiss 473484, step size 0.25 µm) with a low-power objective (Zeiss LD-Epiplan SM ×16, NA 0.30) and a small photometric aperture (0.08 mm equivalent to 3.3 µm on the sample).

The most immediately obvious feature, whereby one trace is offset by 2 µm relative to the other, was caused by movement of the fiber. This was observed progressively throughout the experiment (from pH 5 to pH 8, in steps of 0.5) and originated from loosening of the knots by which the fiber was stretched across the metal sample frame.

The edges (Figure 12.6, 3 and 4) of the fiber are indicated by high absorbance peaks that do not change with pH. In scanning across a cylindrical structure, the edge of the sample has the minimum depth, while the maximum depth is halfway across. But the muscle fiber has a higher refractive index than the surrounding buffer, which

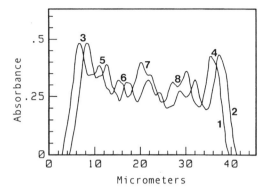

FIGURE 12.6

Scans across a cylindrical muscle fiber at pH 5 (1) and 6.5 (2), showing absorbance peaks at the edges of the fiber (3 and 4) caused by refraction and absorbance peaks, sometimes dented, caused by four A bands (5 to 8).

creates the refractive effect of a bright halo outside a dark rim (depending on the plane of focus) and creates the sharp edges (Figure 12.6, 3 and 4). This effect is often called the Becke line (Slayter, 1970). If sample structures have a higher refractive index than their surrounding medium, the Becke line moves inwards towards the structure when the objective is raised and outwards when the objective is lowered. The muscle fiber is almost round in cross-section, so that it functions as a convex lens, creating a strong Becke line similar to that used in Heyn's method for measuring the refractive index of textile fibers (Morton and Hearle, 1975). Changing the pH caused only a small change in refractive index of the buffer and little or no change in the edge.

Scanning across the fiber, four A bands with a high refractive index were detected (Figure 12.6, 6 to 8), and their absorbance was inversely proportional to pH. As pH went up, absorbance went down, in agreement with the changes in path difference for polarized light described in Chapter 7; however, one must always be alert for other possibilities. For example, the circulating fluid might wash out or dissolve a component of the fiber (such as myosin when using a 0.2-M phosphate buffer), independently of the pH of the buffer, thus causing a time-related decrease in absorbance. Thus, for the experiment in Figure 12.6, the sample was flushed for several hours with ascending and descending buffer before being measured, and the pH-effect also was shown to be reversible (as pH was decreased, absorbance increased). The dimple in the A-band peak (Figure 12.6, 8) was caused by the H zone, between the ends of the thin filaments of the sarcomere.

In conclusion, it is relatively simple to build apparatus to control sample pH for experiments with a CAM. The A and I bands of skeletal muscle demonstrated in Figure 12.6 are certainly not at the highest resolution of the LM but, nevertheless, are well into the working range of normal light microscopy. Thus, the effects of sample chamber depth on resolution are only a partial limitation to this methodology. Doubtless, there are many other ways to control pH apart from that described here.

Also, other software principles, such as fuzzy logic, might be faster; however, speed is relatively unimportant when any cautious investigator would prefer to wait to be sure that the sample has equilibrated to its new programmed pH.

12.3   Refractive Index Control

The refractive index of the medium in the sample chamber (Figure 12.1, 5) may be programmed using a liquid chromatography refractometer; however, in most of the refractometers borrowed from the chromatography laboratory, flow rates through chromatography microtubing will be too slow. High flow-rate refractometers are manufactured but they seem to be quite rare. Another point to watch out for, although it may be strenuously denied by chromatographers themselves, is linearity. Being designed to detect transient peaks in a passing liquid, some chromatography refractometers seem to perform very poorly with a constant high refractive index plateau as will occur in programming the refractive index for a CAM sample chamber.

To increase refractive index, the solenoid valve is opened to a reservoir of strong sucrose solution. To decrease refractive index, the valve to a distilled water or pH buffer reservoir is opened. The high viscosity of concentrated sucrose solution reduces the cycling time of the apparatus and the equilibration time of the sample quite noticeably. The sucrose reservoir needs an additive such as camphor to deter microbial growth, and it is essential to flush the apparatus with water after completion of a day's work; otherwise, crystalline sucrose will clog the solenoid valves. The refractive index of the surrounding medium has major effects on the measurement of birefringence in fibers (Thetford and Simmens, 1969).

12.4   Temperature Control

Temperature-controlled or thermal stages may be constructed from scratch (Mersey et al., 1978; Reid, 1978) or purchased as an accessory for the LM, and they have a variety of applications. Temperature may be decreased until crystallization occurs to measure osmotic pressure or to separate different types of crystals (Diller, 1982; Hsu et al., 1996). A thermal stage also may be used to investigate supercooling in microorganisms (Wharton and Roland, 1984), or to investigate fluorescence quenching (Tiffe and Hundeshagen, 1982). Heating of proteins may be used to examine changes in tertiary structure.

There are many ways in which a thermal stage can be interfaced with a controller:

(1)   Control via GPIB

(2)   Control via RS-232 (this is less desirable than control via GPIB because of competition for this limited-resource bus)

(3)   Control by an applied voltage generated from a DAC

(4)   Control by an applied resistance (however, a resistance output ties up numerous relays and requires precision resistors to form a programmable decade sequence)

(5)   Control by an applied voltage or resistance tapped into the manufacturer's control bridge (this is difficult if the front-panel knob rotates ganged potentiometers)

(6)   Control by a knob-turning actuator, as described in Chapter 7

12.4.1   Temperature Sensor

A thermocouple has two dissimilar electrical conductors joined at their ends and generates a thermoelectric voltage following the temperature difference between the two end junctions. Iron-constantan and chromel-alumel thermocouples are common examples. Junctions may be covered or exposed, and with or without grounding. Resistive circuit components (thermistors) may come as a two-terminal semiconductor with resistance decreasing as temperature is increased. A Callendar's thermometer or platinum RTD probe has a thin film or wire of pure platinum which, for a metal, has a relatively high resistance. Thus, temperature may be measured as an electrical resistance.

The time constant is the length of time that the sensor takes to read the true temperature of the sample. The major problem is that, with a microscopic sample, heat conduction along the sensor, its wire connections, and its protective shielding may alter the temperature of the sample. Thus, for temperature measurement on the LM stage, the sample generally is equilibrated to its medium, then the temperature of the medium is measured. When making remote measurements, care is needed with the length of wire between the sensor and the controller. If it exceeds the manufacturer's specifications, a signal transmitter may be required.

Finally, the whole CAM itself may be used as a temperature sensor for a special application, such as monitoring microwave heating, where metallic components cannot be used. A quartz optical fiber is fitted with a temperature-sensitive fluorescent cap, and temperature is measured as a change in fluorescence emission, using fluorometry principles covered in Chapter 11.

12.4.2   Thermal Stage

Usually the thermal stage is placed on top of the existing LM or polarizing microscope stage (Moghimi et al., 1996), then the microscope slide containing the sample is placed on the thermal stage. But raising the underneath of the slide beyond the normal top limit of the substage condenser may create a problem for working at high magnification. Thus, the upper limit stop of the substage condenser rack may need adjustment. If this is done, however, it is important to remember to rack down the condenser whenever the thermal stage is moved laterally; otherwise, it may collide with the top of the condenser.

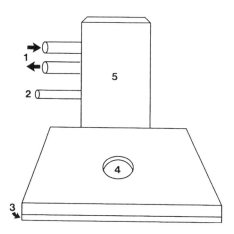

FIGURE 12.7

Thermal stage, with in and out water supplies (1), electrical connections (2), heat insulation from the LM stage (3), optical aperture (4), and heat pump stack (5).

A typical thermal stage, such as the Sensortek TS-4 is shown in Figure 12.7. In and out water supplies (Figure 12.7, 1) are required for the thermoelectric heat pump, keeping one junction at a constant temperature. Being relatively stiff Tygon tubes, they may create problems if the thermal stage is mounted on a scanning stage (causing an unanticipated load on the stepper motor and gear train). A power supply is required for the heat pump (Figure 12.7, 5), as well as heat insulation against the LM stage (Figure 12.7, 3). Although adequate for maintaining a fixed temperature, the weakness of this arrangement may become apparent in programming variable temperatures. Normally, the whole microscope slide is maintained at a fixed temperature, and heat transfer through the glass is sufficient to reach across the optical aperture (Figure 12.7, 4). Thus, the sample in the axis of the aperture is probably at, or close to, the temperature of the remainder of the slide, especially when the slide is hot. A powerful CAM illuminator may radiate an appreciable amount of heat, even if the beam has passed through a heat filter. This heat may be unimportant if the stage is hot but may cause problems if the sample is supposed to be frozen. The front face of the heat pump at the back of the stage (Figure 12.7, 5) provides a down-draft of cold air over the slide which may help. It is important to have an independent temperature sensor on the slide, just slightly to one side of the photometric aperture. This, rather than the built-in sensor of the thermal stage, should be used as the control sensor and to measure the temperature of the sample when data are collected. If heat transfer problems are encountered, it may be necessary to bias the programmed temperature for the thermal stage in order to attain an actual temperature for the sample. For example, to reach a sample temperature of 0°C, it may be necessary to program the thermal stage to −5°C.

The main problem when working with a thermal stage at low temperatures is the occurrence of condensation on the microscope slide or objective. This may alter or invalidate optical measurements made with the CAM.

12.4.3   Control Loops

Several strategies are possible to avoid control-loop oscillations:

(1)   The hard-wired bridge controller of the thermal stage may be used to maintain temperature, adjusting the balance point of the bridge with the PC; however, this may make it difficult to use an external temperature sensor located on the glass next to the measured area of the sample.

(2)   A slave digital controller on the GPIB may be used, mediating directly between a sensor located on the sample and the control circuit for the heat pump.

(3)   The main controller may be used, as in option 2, making an optical measurement immediately once the programmed temperature has been reached.

Regardless of which strategy is used, the temperature to be logged in the data set should not be the nominal programmed temperature but rather an average of temperature measurements made immediately before and after the optical measurement. In most cases, this will cancel any systematic errors which may arise from temperature oscillations and drifting.

12.5   Planning the Experimental Protocol

User-friendly software requires the anticipation of user requirements. A one-way measuring sequence is simple to plan, either taking the sample from the minimum value to the maximum value or vice versa. An example of a relatively simple destructive heat test is shown in Figure 12.8, where unfolding of the triple helix of tropocollagen in a heated collagen fiber was detected between 70 and 75°C by loss of birefringence. Note the irregularities in temperature incrementation revealed by using measured values rather than nominal programmed values.

Some experiments may require a complete cycle of measurements, returning to the starting point. For example, starting at a room temperature of 20°C, the sample is to be cooled to 10°, heated to 30°, then brought back to 20°C. The measuring loop may be programmed as follows:

```
10    Plan:SUB Plan ! plan measuring loop
20    ALLOCATE Seq(10)! dummy sequence
30    ALLOCATE Revseq(10)! dummy reverse sequence
40    PRINT "enter current value", ! or call
      measuring routine
```

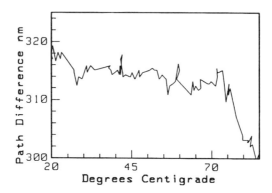

FIGURE 12.8
Loss of birefringence in a perimysial collagen fiber heated with a thermal stage.

```
50      INPUT  Now
60      PRINT  Now
70      PRINT  "enter  increment";
80      INPUT  Inc
90      PRINT  Inc
100     PRINT  "enter  maximum";
110     INPUT  Max
120     PRINT  Max
130     PRINT  "enter  minimum";
140     INPUT  Min
150     PRINT  Min
160     PRINT
170     PRINT  "for  descending  cycle...enter  0"
180     PRINT  "for  ascending  cycle..........1"
190     PRINT  "for  descending  gradient......2"
200     PRINT  "for  ascending  gradient.......3"
210     INPUT  Dir
220     !
230     !......................
240     !
250     !  set-up  matrices  and  sequence
260     !
270     IF  Dir<2  THEN  Number=(((Max-Min)/Inc)*2)! base  0
280     IF  Dir>=2  THEN  Number=((Max-Min)/Inc)! base  0
290     DEALLOCATE  Seq(*),Revseq(*)
300     ALLOCATE  Seq(Number)!  theoretical  sequence
310     ALLOCATE  Revseq(Number)!  reverse  Seq
320     !
330     !......................
340     !
350     !  plan  loops  and  one-ways
360     !
```

```
370    IF Dir<2 THEN ! is a loop cycle
380    IF Now>Min AND Now<Max AND Dir=0 THEN
       Seq(0)=Now-(Now MOD          Inc)
390    IF Now>Min AND Now<Max AND Dir=1 THEN
       Seq(0)=Now-(Now MOD Inc)+Inc
400    IF Now>=Max THEN
410    Seq(0)=Max
420    Dir=0
430    END IF
440    IF Now<=Min THEN
450    Seq(0)=Min
460    Dir=1
470    END IF
480    FOR I=1 TO Number
490    IF Dir=1 AND Seq(I-1)>=Max THEN Dir=0
500    IF Dir=0 AND Seq(I-1)<=Min THEN Dir=1
510    IF Dir=1 THEN Seq(I)=Seq(I-1)+Inc
520    IF Dir=0 THEN Seq(I)=Seq(I-1)-Inc
530    NEXT I
540    END IF
550    IF Dir=2 THEN ! descending one way only
560    FOR I=0 TO Number
570    Seq(I)=Max-(I*Inc)
580    NEXT I
590    END IF
600    IF Dir=3 THEN ! ascending one way only
610    FOR I=0 TO Number
620    Seq(I)=Min+(I*Inc)
630    NEXT I
640    END IF
650    PRINT "Sequence to measure is ",Seq(*)
660    SUBEND
```

The result of running the subprogram is as follows:

```
enter current value 20
enter increment 1
enter maximum 30
enter minimum 10

for descending cycle...enter 0 (this option selected)
for ascending cycle..........1
for descending gradient......2
for ascending gradient.......3

Sequence to measure is  20 19 18 17 16 15 14 13 12
      11 10 11 12 13 14 15 16 17 18 19 20 21 22 23
      24 25 26 27 28 29 30 29 28 27 26 25 24 23 22
      21 20
```

In cyclic testing, if an optical measurement is made when the state of the sample reaches or exceeds a programmed value, a logical direction pointer must be maintained; otherwise, the algorithm will not wait on the descending cycle until the sample has cooled to the programmed temperature. Thus, on the way up and at the maximum, a temperature ≥ programmed value is accepted, while on the way down or at the minimum, temperature ≤ programmed value is accepted. A window of acceptance might be used, but then it is difficult to program the overall rate of temperature change, which usually is critical in biological experiments in which experimental effects may depend on a time-temperature interaction. Unless the window of acceptance is large (in which case ascending and descending temperatures will not match), the controller may oscillate around the window, making it impossible to achieve the programmed rate of change.

References

Adamovsky, G. and N. D. Piltch. 1986. Fiber-optic thermometer using temperature dependent absorption, broadband detection, and time domain referencing. *Appl. Opt.* 25: 4439–4443.

Alabbas, S. H., D. C. Ashworth, B. Bezzaa, S. A. Momin, and R. Narayanaswamy. 1996. Factors affecting the response time of an optical-fibre reflectance pH sensor. *Sensors Actuators.* A51: 129–134.

Allen, R. D., J. W. Brault, and R. M. Zeh. 1966. Image contrast and phase-modulated light methods in polarisation and interference microscopy. In: R. Barer and V. E. Cosslett (Eds.), *Advances in Optical and Electron Microscopy.* Vol. 1. Academic Press, London, pp. 77–114.

Ames, N. P., R. D. Hartley, and D. E. Akin. 1992. Distribution of aromatic compounds in coastal Bermudagrass cell walls using ultraviolet absorption scanning microdensitometry. *Food Struct.* 11: 25–32.

Andrejevic, S., J. F. Savary, P. Monnier, C. Fontolliet, D. Braichotte, G. Wagnieres, and H. Vandenbergh. 1996. Measurements by fluorescence microscopy of the time-dependent distribution of meso-tetra-hydroxyphenylchlorin in healthy tissues and chemically induced "early" squamous cell carcinoma of the Syrian hamster cheek pouch. *J. Photochem. Photobiol.* 36: 143–151.

Arimoto, R. and J. M. Murray. 1996. Orientation-dependent visibility of long thin objects in polarization-based microscopy. *Biophys. J.* 70: 2969–2980.

Arokoski, J. P. A., M. M. Hyttinen, T. Lapvetelainen, P. Takacs, B. Kosztbaczky, L. Modis, V. Kovanen, and H. J. Helminen. 1996. Decreased birefringence of the superficial zone collagen network in the canine knee (stifle) articular cartilage after long distance running training, detected by quantitative polarised light microscopy. *Ann. Rheumat. Dis.* 55: 253–264.

Auffermann, W., S. T. Wu, W. W. Parmley, and J. Wikman-Coffet. 1990. Glycolysis in heart failure: a ^{31}P-NMR and surface fluorometry study. *Bas. Res. Cardiol.* 85: 342–357.

Bashford, C. L. 1987. An introduction to spectrophotometry and fluorescence spectrometry. In: D. A. Harris and C. L. Bashford (Eds.), *Spectrophotometry and Spectrofluorimetry.* IRL Press, Oxford, p. 15.

Beach, J. M., E. D. McGahren, J. Xia, and B. R. Duling. 1996. Ratiometric measurement of endothelial depolarization in arterioles with a potential-sensitive dye. *Am. J. Physiol.* 39: H2216–H2227.

Bellmunt, M. J., M. Portero, R. Pamplona, L. Cosso, P. Odetti, and J. Prat. 1995. Evidence for the Maillard reaction in rat lung collagen and its relationship with solubility and age. *Biochem. Biophys. Acta.* 1272: 53–60.

Bellon, G., A. Randoux, and J-P. Borel. 1985. A study of collagen metabolism in cell cultures by fluorometric determination of proline and hydroxyproline. *Collagen Rel. Res.* 5: 423–435.

Benaim, N., K. T. V Grattan, and A. W. Palmer. 1986. Simple fibre optic pH sensor for use in liquid titrations. *Analyst* 111: 1095–1097.

Bennett, H. S. 1950. Methods applicable to the study of both fresh and fixed material. The microscopical investigation of biological materials with polarized light. In: R. McClung Jones (Ed.), *McClung's Handbook of Microscopical Technique*, 3rd ed. Paul Hoeber, New York, pp. 591–677.

Berthold, C-H., J-O. Kellerth, and S. Conradi. 1979. Electron microscopic studies of serially sectioned cat spinal α-motoneurons. I. Effects of microelectrode impalement and intracellular staining with the fluorescent dye "procion yellow". *J. Comp. Neurol.* 184: 709–740.

Birth, G. S., C. E. Davis, and W. E. Townsend. 1978. The scatter coefficient as a measure of pork quality. *J. Anim. Sci.* 46: 639–645.

Blum, L. J., S. M. Gautier, and P. R. Coulet. 1988. Luminescent fiber-optic biosensor. *Anal. Lett.* 21: 717–726.

Boguth, W. and H. Piller. 1988. The influence of the aperture of illumination on the accuracy of microscope-photometric measurements of internal transmittance. *J. Microsc.* 149: 117–125.

Boppart, S. A., M. E. Brezinski, B. B. Bouma, G. J. Tearney, and J. G. Fujimoto. 1996. Investigation of developing embryonic morphology using optical coherence tomography. *Develop. Biol.* 177: 54–63.

Bornfleth, H., K. Aldinger, M. Hausmann, A. Jauch, and C. Cremer. 1996. Comparative genomic hybridization imaging by the one-chip true-color CCD camera Kappa CF 15 MC. *Cytometry.* 24: 1–13.

Bowen, W. J. 1949. The absorption spectra and extinction coefficients of myoglobin. *J. Biol. Chem.* 179: 235–245.

Bracegirdle, B. 1978. *A History of Microtechnique.* Cornell University Press, Ithaca, NY.

Braga, P. C. 1990. A variable-thickness multipurpose culture chamber for high-magnification observation. *J. Microsc.* 159: 285–288.

Bretagnon, T., K. Abdurahman, D. Kerr, and S. Dannefaer. 1992. Beyond 10^6-count lifetime spectra using Hamamatsu tubes and Pilot U. *Mater. Sci. Forum (Switz.).* 105–110: 1841–1844.

Bright, D. S. and E. B. Steel. 1987. Two-dimensional top hat filter for extracting spots and spheres from digital images. *J. Microsc.* 146: 191–200.

Butler, W. L. 1962. Absorption of light by turbid materials. *J. Opt. Soc. Am.* 52: 292–299.

Chen, P. C., T. H. Lin, W. L. Wu, and J. L. Wu. 1994. *Multidimensional Microscopy.* Springer-Verlag, New York.

Chen, H., D. D. Hughes, T. A. Chan, J. W. Sedat, and D. A. Agard. 1996. IVE (image visualization environment): a software platform for all three-dimensional microscopy applications. *J. Struct. Biol.* 116: 56–60.

Chien, J. C. W. and E. P. Chang. 1972. Dynamic mechanical and rheo-optical studies of collagen and elastin. *Biopolymers.* 11: 2015–2031.

Chikamori, K. and M. Yamada. 1986. Determination of dehydrogenase activity in tissue sections by tridensitometry and its applications. *Acta Histochem. Cytochem.* 19: 41–50.

Chowdhury, T. K. 1969. Techniques of intracellular microinjection. In: M. Lavalee, O. F. Schanne, and N. C. Hebert (Eds.), *Glass Microelectrodes.* John Wiley, New York, pp. 404–423.

Clokey, G. V. and L. A. Jacobson. 1986. The autofluorescent "lipofuscin granules" in the intestinal cells of *Caenorhabditis elegans* are secondary lysosomes. *Mechanisms Ageing Develop.* 35: 79–94.

Conway, J. M., K. H. Norriss, and C. E. Bodwell. 1984. A new approach for the estimation of body composition: infrared interactance. *Am. J. Clin. Nutr.* 40: 1132–1130.

Cooper, T. G. 1977. *The Tools of Biochemistry*, John Wiley, New York.

Cornelese-ten Velde, I., J. Bonnet, H. J. Tanke, and J. S. Ploem. 1990. Reflection contrast microscopy performed on epi-illumination microscope stands: comparison of reflection contrast- and epi-polarization microscopy. *J. Microsc.* 159: 1–13.

Cowden, R. R. and S. K. Curtis. 1975. A comparison of four quantitative cytochemical methods directed towards demonstration of DNA. *Histochemistry.* 45: 299–308.

Crissman, H. A., M. S. Oka, and J. A. Steinkamp. 1976. Rapid staining methods for analysis of deoxyribonucleic acid and protein in mammalian cells. *J. Histochem. Cytochem.* 24: 64–71.

Daly, J. C. 1984. *Fiber Optics.* CRC Press, Boca Raton, FL.

David, G. B. and W. Galbraith. 1974. The Denver universal microspectroradiometer (DUM). I. General design and construction. *J. Microsc.* 103: 135–178.

Davidson, R. S. 1996. The photodegradation of some naturally occurring polymers. *J. Photochem. Photobiol. B.* 33: 3–25.

Davis, A. and F. J. Vastola. 1977. Developments in automated reflectance microscopy of coal. *J. Microsc.* 109: 3–12.

Davis, A., K. W. Kuehn, D. H. Maylotte, and R. L. St. Peter. 1983. Mapping of polished coal surfaces by automated reflectance microscopy. *J. Microsc.* 132: 297–302.

Davydov, D. R., T. V. Knyushko, I. P. Kanaeva, Y. M. Koen, N. F. Samenkova, A. I. Archakov, and G. H. B. Hoa. 1996. Interactions of cytochrome P450 2B4 with DADPH-cytochrome P450 reductase studied by fluorescent probe. *Biochimie.* 78: 8–9.

De Sénarmont, H. 1840. Sur les modifications que la réflexion spéculaire à la surface des corps métalliques imprime à un rayon de lumière polarisée. *Ann. Chim. Phys.* 73: 337–362.

Decarvalho, H. F. and S. R. Taboga. 1996. Fluorescence and confocal laser scanning microscopy imaging of elastic fibers in hematoxylin-eosin stained sections. *Histochem. Cell Biol.* 106: 587–592.

Decarvalho, H. F. and B. D. Vidal. 1996. Polarized microscopy study of an antennal sensillum of *Triatoma infestans*: an ordered distribution of chitin fibrils and associated components. *Comput. Roy. Acad. Sci., Ser. III.* 319: 33–38.

Delpy, D. T., M. C. Cope, E. B. Cady, J. S. Wyatt, P. A. Hamilton, P. L. Hope, S. W. Ray, and E. O. R. Reynolds. 1987. Cerebral monitoring in newborn infants by magnetic resonance and near infrared spectroscopy. *Scand. J. Clin. Lab. Invest.* 47(Suppl. 188): 9–17.

Devlin, T., J. Cruz, U. Joshi, K. Kazlauskis, C. Muehleisen, and T. S. Yang. 1988. Phototube testing for CDF. *Nucl. Instrum. Methods Phys. Res. A, Accel. Spectrom. Detect. Assoc. Equip. (Neth.)* A268: 24–32.

Diller, K. R. 1982. Quantitative low temperature optical microscopy of biological systems. *J. Microsc.* 126: 9–28.

Dunn, G. A. 1991. Quantitative interference microscopy. In: R. C. Cherry (Ed.), *New Techniques of Optical Microscopy and Microspectrophotometry*. CRC Press, Boca Raton, FL, pp. 91–118.

Dvorak, J. A., T. R. Clem, and W. F. Stotler. 1971. The design and construction of a computer-compatible system to measure and record optical retardation with a polarizing or interference microscope. *J. Microsc.* 96: 109–114.

Ebrahimzadeh, P. R., F. Bazagani, F. Afzal, C. Hogfors, and M. Braide. 1996. A subpopulation analysis of f-MLP stimulated granulocytes migrating in filters. *Biorheology.* 33: 231–150.

Engelmann, T. W. 1878. Neue Untersuchungen über die mikroskopischen Vorgänge bei der Muskelcontraktion. *Pflügers Arch.* 18: 1–25.

Ermakov, V. I., A. V. Komotskov, and I. S. Moshnikov. 1986. Photodetector with electronic protection from strong exposures. *Instrum. Exp. Tech. (USA).* 29: 246–247.

Finch, S. A. E. and A. Stier. 1988. A perfusion chamber for high-resolution light microscopy of cultured cells. *J. Microsc.* 151: 71–75.

Fischer, J-G., H. Mewes, H-H. Hopp, and R. Schubert. 1996. Analysis of pressurized resistance vessel diameter changes with a low cost digital image processing device. *Comput. Methods Programs Biomed.* 50: 23–30.

Fischmeister, H. F. 1968. Scanning methods in quantitative metallography. In: D. T. DeHoff and F. N. Rhines (Eds.), *Quantitative Microscopy*. McGraw-Hill, New York, pp. 336–379.

Fitch, P. and A. G. Gargus. 1985. Remote UV-VIS-NIR spectroscopy using fiber optic chemical sensing. *Amer. Lab.* 17(12): 64–69.

Florijn, R. J., J. Bonnet, H. Vrolijk, A. K. Raap, and H. J. Tanke. 1996. Effect of chromatic errors in microscopy on the visualization of multi-colored fluorescence in situ hybridization. *Cytometry* 23: 8–14.

Frohlich, M. W. 1986. Birefringent objects visualized by circular polarization microscopy. *Stain Technol.* 61: 139–143.

Fujii, Y. and Y. Yamazaki. 1990. A fibre-optic 3-D microscope with high depth sensitivity. *J. Microsc.* 158: 145–151.

Galassi, L. and B. Della Vecchia. 1988. Selection of optimal transmittances with the cytophotometric two-wavelength method. *Bas. Appl. Histochem.* 32: 279–291.

Galbraith, W., S. B. Geyer, and G. B. David. 1975. The Denver universal microspectroradiometer (DUM). II. Computer configuration and modular programming for radiometry. *J. Microsc.* 105: 237–264.

Gibson, K. S. 1949. Spectrophotometry. In: *U.S. Department of Commerce, National Bureau of Standards*, Circular 484.

Goldstein, D. J. 1969. Detection of dichroism with the microscope. *J. Microsc.* 89: 19–36.

Goldstein, D. J. 1975. Aspects of scanning microdensitometry. III. The monochromator system. *J. Microsc.* 105: 33–56.

Goranson, R. W. and L. H. Adams. 1933. A method for the precise measurement of optical path-difference, especially in stressed glass. *J. Franklin Inst.* 216: 475–504.

Gurrieri, S., S. B. Smith, K. S. Wells, I. D. Johnson, and C. Bustamante. 1996. Real-time imaging of the re-orientation mechanisms of YOYO-labelled DNA molecules during 90 degrees and 120 degrees pulsed field gel electrophoresis. *Nucleic Acids Res.* 24: 4759–4767.

Hammersen, F. and K. Messmer. 1984. *Skeletal Muscle Microcirculation*. S. Karger, Basel.

Hartley, W. G. 1993. *The Light Microscope. Its Use and Development*. Senecio Publishing, Oxford.

Hartshorne, N. H. and A. Stuart. 1970. *Crystals and the Polarising Microscope*, 4th ed. Edward Arnold, London, pp. 308–318.

Hartveit, E. 1996. Membrane currents evoked by ionotropic glutamate receptor agonists in rod bipolar cells in the rat retinal slice preparation. *J. Neurophysiol.* 76: 401–422.

Hecht, E. 1987. *Optics*. Addison-Wesley, Reading, MA.

Heilbronner, R. P. 1988. Distortion of orientation data introduced by digitizer procedures. *J. Microsc.* 149: 83–96.

Heldal, K., A. Skogstad, and W. Eduard. 1996. Improvements in the quantification of airborne micro-organisms in the farm environment by epifluorescence microscopy. *Ann. Occupat. Hyg.* 40: 437–447.

Hewlett-Packard. 1982. *Digital Data Transmission with the HP Fiber Optic System*. Application Note 1000. Hewlett-Packard, Palo Alto, CA.

Hirshberg, A., A. Buchner, and D. Dayan. 1996. The central odontogenic fibroma and the hyperplastic dental follicle: study with picrosirius red and polarizing microscopy. *J. Oral Pathol. Med.* 25: 125–127.

Hsu, C. C., A. J. Walsh, H. M. Nguyen, D. E. Overcashier, H. Koningbastiaan, R. C. Bailey, and S. J. Nail. 1996. Design and application of a low-temperature Peltier-cooling microscope stage. *J. Pharm. Sci.* 85: 70–74.

Huether, G. and V. Neuhoff. 1981. Microelectrophoresis as a tool in enzyme histochemistry. *Histochem. J.* 13: 207–225.

Hurlbut, C. S. 1959. *Dana's Manual of Mineralogy*, 18th ed. John Wiley, New York, pp. 156–157.

Huxley, A. 1980. *Reflections on Muscle*. Princeton University Press, Princeton, NJ.

Huxley, E. E. and J. Hanson. 1957. Quantitative studies on the structure of cross-striated myofibrils. I. Investigations by interference microscopy. *Biochem. Biophys. Acta.* 23: 229–249.

Huxley, A. F. and R. Niedergerke. 1954. Structural changes in muscle during contraction. Interference microscopy of living muscle fibres. *Nature.* 173: 971–973.

Inoué, S. 1986. *Video Microscopy*. Plenum Press, New York.

Johnson, F. A. 1970. Investigations of source-dependent contributions to photomultiplier noise. *Nucl. Instr. Meth.* 87: 215–220.

Jones, H. M., R. J. Baskin, and Y. Yeh. 1991. The molecular origin of birefringence in skeletal muscle. Contribution of myosin subfragment S-1. *Biophys. J.* 60: 1217–1228.

Judd, D. B. and G. Wyszecki. 1975. *Color in Business, Science and Industry*. John Wiley, New York.

Kapany, N. S. 1967. *Fiber Optics: Principles and Applications*. Academic Press, New York.

Kawai, H., S. Nakamura, M. Mimuro, M. Furuya, and M. Watanabe. 1996. Microspectrophotometry of the autofluorescent flagellum in phototactic brown algal zoids. *Protoplasma.* 191: 172–177.

Kenny, L. C. 1983. Automated focusing of an optical microscope. *J. Microsc.* 132: 97–107.

King, R. J. 1957. The optical correction of objectives used for microspectroscopy. *J. Roy. Microsc. Soc.* 77: 99–106.

Knoll, G. F. 1979. *Radiation Detection and Measurement*. John Wiley, New York, pp. 272–305.

Krammer, B. and K. Uberriegler. 1996. In-vitro investigation of ALA-induced protoporphyrin IX. *J. Photochem. Photobiol.* 36: 121–126.

Kunz, W. S. and F. N. Gellerich. 1993. Quantification of the content of fluorescent flavoproteins in mitochondria from liver, kidney cortex, skeletal muscle, and brain. *Biochem. Med. Metab. Biol.* 50: 103–110.

Lansing Taylor, D. and R. M. Yeh. 1976. Methods for the measurement of polarization optical properties. I. Birefringence. *J. Microsc.* 108: 251–259.

Laughlin, R. G., A. M. Marrer, C. Marcott, and R. L. Munyon. 1985. An improved Sénarmont retardation analysis method. *J. Microsc.* 139: 239–247.

Leggett, K. 1997. Fiber optic sensor targets biological culprits. *Biophotonics.* 4(1): 55–55.

Lewis, A. 1991. The confluence of advances in light microscopy: CCD, confocal, near-field and molecular exciton microscopy. In: R. C. Cherry (Ed.), *New Techniques of Optical Microscopy and Microspectrophotometry*. CRC Press, Boca Raton, FL, pp. 49–89.

Loew, L. M. 1991. Membrane potential imaging. In: R. C. Cherry (Ed.), *New Techniques of Optical Microscopy and Microspectrophotometry*. CRC Press, Boca Raton, FL, pp. 255–272.

Lundgren, L., S. Lundstrom, G. Sundstrom, G. Bergman, and S. Krantz. 1996. A quantitative method using a light microscope on-line to a Macintosh computer for the analysis of tremolite fibres in dolomite. *Ann. Occupat. Hyg.* 40: 197.

Mannheimer, W. A. 1996. BibMic: a bibliography of books relating to materials microscopy. *Materials Charact.* 36: 105–149.

Marchant, J. L. and J. M. V. Blanshard. 1978. Studies on the dynamics of the gelatinization of starch granules employing a small angle light scattering system. *Stärke*. 30, 257–264.

Mayer, R. T. and V. M. Novacek. 1974. A direct recording corrected microspectrofluorometer. *J. Microsc.* 102: 165–177.

Mayevsky, A. 1984. Brain NADH redox state monitored in vivo by fiber optic surface fluorometry. *Brain Res. Rev.* 7, 49–68.

McFarland, D. C., J. E. Pesall, K. K. Gilkerson, and A. S. Todd. 1993. Comparison of the proliferation and differentiation of myogenic satellite cells derived from Merriam's and commercial varieties of turkeys. *Comp. Biochem. Physiol.* 104A: 455–460.

McGrath, J. J. 1985. A microscope diffusion chamber for the determination of the equilibrium and non-equilibrium osmotic response of individual cells. *J. Microsc.* 139: 249–263.

Mendelsohn, M. L. 1966. Absorption cytophotometry: comparative methodology for heterogeneous objects, and the two-wavelength method. In: G. L. Wied (Ed.), *Introduction to Quantitative Cytochemistry*. Academic Press, New York, pp. 201–214.

Mercer, R. R., S. H. Randell, and S. L. Young. 1990. Measurement of boundaries using a digitizer tablet. *J. Microsc.* 160: 97–105.

Mersey, B., M. E. McCully, and E. Fatica. 1978. An inexpensive controlled temperature stage which allows high resolution optical microscopy. *J. Microsc.* 113: 307–310.

Milanovich, F. P., P. F. Daley, S. M. Klainer, and L. Eccles. 1986. Remote measurement of organochlorides with a fiber optic sensor. II. A dedicated portable fluorimeter. *Anal. Instrum.* 15: 347–358.

Miller, E. J. 1985. Recent information on the chemistry of the collagens. In: W. T. Butler (Ed.), *The Chemistry and Biology of Mineralized Tissues*. EBSCO Media, Birmingham, England.

Moghimi, H. R., A. C. Williams, and B. W. Barry. 1996. A lamellar matrix model for stratum corneum intercellular lipids. 1. Characterisation and comparison with stratum corneum intercellular structure. *Int. J. Pharmaceut.* 131: 103–115.

Morgan, C. G. and A. C. Mitchell. 1996. Fluorescence lifetime imaging: an emerging technique in fluorescence microscopy. *Chromosome Res.* 4: 261–263.

Morrison, W. H., D. E. Akin, G. Ramaswamy, and B. Baldwin. 1996. Evaluating chemically retted kenaf using chemical, histochemical, and microspectrophotometric analyses. *Textile Res. J.* 66: 651–656.

Morton, W. E. and J. W. S. Hearle. 1975. *Physical Properties of Textile Fibres*. Heinemann, London, pp. 567–570.

Mozdziak, P. E., T. Fassel, R. Gregory, E. Schultz, M. L. Greaser, and R. G. Cassens. 1994. Quantitation of satellite cell proliferation in vivo using image analysis. *Biotech. Histochem.* 69: 249–252.

Na, G. C. 1988. UV spectroscopic characterization of type I collagen. *Collagen Rel. Res.* 8: 315–330.

Naughton, J. J., M. M. Frodyma, and H. Zeitlin. 1957. Spectral reflectance applied to the study of heme pigments. *Science.* 125: 125.

Neti, S., T. J. Butrie, and J. C. Chen. 1986. Fiber-optic liquid contact measurements in pool boiling. *Rev. Sci. Instrum.* 57: 3043–3047.

Nickalls, R. W. D. and R. Ramasubramanian. 1995. *Interfacing the IBM-PC to Medical Equipment: The Art of Serial Communication.* Cambridge University Press, London.

Nollie, G. J., H. S. Sandhu, Z. Z. Cernovsky, and P. B. Canham. 1996. Regional differences in molecular cross-linking of periodontal ligament collagen of rat incisor by polarizing microscopy. *Connect. Tissue Res.* 33: 283–289.

Nolte, J. and D. Pette. 1972. Microphotometric determination of enzyme activity in single cells in cryostat sections. I. Application of the gel film technique to microphotometry and studies on the intralobular distribution of succinate dehydrogenase and lactate dehydrogenase activities in rat liver. *J. Histochem. Cytochem.* 20: 567–576.

Nys, B., W. Jacob, and P. Van Espen. 1991. The use of low-cost frame grabber boards on PC-AT computers for ESI: possibilities and limitations. *J. Microsc.* 162: 55–60.

Odetti, P., M. A. Pronzato, G. Noberasco, L. Cosso, N. Traverso, D. Cottalasso, and U. M. Marinari. 1994. Relationship between glycation and oxidation related fluorescences in rat collagen during aging. An in vivo and in vitro study. *Lab. Invest.* 70: 61–67.

Oldenbourg, R. 1996. A new view on polarization microscopy. *Nature* 381: 811–812.

Ong, S. H., X. C. Jin, and J. R. Sinniah. 1996. Image analysis of tissue sections. *Comput. Biol. Med.* 26: 269–279.

Ontell, M. 1974. Muscle satellite cells: a validated technique for light microscopic identification and a quantitative study of changes in their population following denervation. *Anat. Rec.* 178: 211–225.

Oostveldt, P. van and G. Boeken. 1976. Absorption cytophotometry: evaluation of three methods for the determination of DNA in Feulgen stained nuclei. *Histochemistry.* 50: 147–159.

Orford, H. and A. Lockett. 1931. *Lens-Work for Amateurs.* Sir Isaac Pitman & Sons, London.

Ortmann, R. 1975. Use of polarized light for quantitative determination of the adjustment of the tangential fibres in articular cartilage. *Anat. Embryol.* 148: 109–120.

Oster, G. 1955. Birefringence and dichroism. In: G. Oster and A. W. Pollister (Eds.), *Physical Techniques in Biological Research*, Vol. 1. Academic Press, New York, pp. 439–460.

Pawley, J. B. 1990. *Handbook of Biological Confocal Microscopy.* Plenum Press, New York.

Pearse, A. G. E. 1972. *Histochemistry: Theoretical and Applied*, Vol. 2. Churchill Livingstone, Edinburgh, pp. 1229–1230.

Pearse, A. G. E. 1968. *Histochemistry: Theoretical and Applied*, Vol. 1. Churchill Livingstone, Edinburgh, pp. 254–264.

Peters, R. and M. Scholz. 1991. Fluorescence photobleaching techniques. In: R. C. Cherry (Ed.), *New Techniques of Optical Microscopy and Microspectrophotometry*. CRC Press, Boca Raton, FL, pp. 199–228.

Peterson, J. I. and G. G. Vurek. 1984. Fiber-optic sensors for biomedical applications. *Science*. 224: 123–127.

Pickering, J. G., C. M. Ford, and L. H. Chow. 1996. Evidence for rapid accumulation and persistently disordered architecture of fibrillar collagen in human coronary restenosis lesions. *Am. J. Cardiol*. 78: 633–637.

Piller, H. 1973. Modern techniques in reflectance measurements. *J. Microsc*. 100: 35–48.

Piller, H. 1977. *Microscope Photometry*. Springer-Verlag, Berlin.

Piller, H. 1979. Domains of microscope photometry in materials science. *J. Microsc*. 116: 295–310.

Piller, H. 1981. Checking of linearity of the photoelectric system of a microscope photometer. *J. Microsc*. 121: 221–224.

Pimentel, E. R. 1981. Form birefringence of collagen bundles. *Acta Histochem. Cytochem*. 14: 35–40.

Pluta, M. 1988. *Advanced Light Microscopy*, Vol. 1, *Principles and Basic Properties*. Elsevier, Amsterdam; Polish Scientific, Warszawa.

Pluta, M. 1989. *Advanced Light Microscopy*, Vol. 2, *Specialized Methods*. Elsevier, Amsterdam; Polish Scientific, Warszawa.

Puchtler, H., F. S. Waldrop, and L. S. Valentine. 1973. Fluorescence microscopic distinction between elastin and collagen. *Histochemie*. 35: 17–30.

Rabau, M. Y., A. Hirshberg, Y. Hiss, and D. Dayan. 1995. Intestinal anastomosis healing in rat: collagen concentration and histochemical characterization by picrosirius red staining and polarizing microscopy. *Exp. Molec. Pathol*. 62: 160–165.

Rasch, E. M. and R. W. Rasch. 1979. Applications of microcomputer technology to cytophotometry. *J. Histochem. Cytochem*. 27: 1384–1387.

Ray, G. B. and G. H. Paff. 1930. A spectrophotometric study of muscle hemoglobin. *Am. J. Physiol*. 94: 521–528.

Reid, D. S. 1978. A programmed controlled temperature microscope stage. *J. Microsc*. 114: 241–248.

Roche, E. J. and R. F. Van Kavelaar. 1989. Photometric measurement of small optical path differences. *J. Microsc*. 157: 181–186.

Romhányi, G. 1986. Specific topo-optical reactions of connective tissue elements and their ultrastructural interpretation. *Connective Tiss. Res*. 15: 13–16.

Rome, E. 1967. Light and X-ray diffraction studies of the filament lattice of glycerol-extracted rabbit psoas muscle. *J. Molec. Biol*. 27: 591–602.

Ross, K. F. A. 1967. *Phase Contrast and Interference Microscopy for Cell Biologists*. Edward Arnold, London.

Rost, F. W. D. 1995. *Fluorescence Microscopy*. Cambridge University Press, London.

Ruch, F. 1951. Eine Apparatur zur Messung des Ultraviolett-dichroismus von Zelstrukturen. *Exp. Cell Res.* 2: 680–683.

Ruch, F. 1966. Birefringence and dichroism of cells and tissues. In: A. W. Pollister (Ed.), *Physical Techniques in Biological Research*, Vol. III, Pt. A. Academic Press, New York, pp. 57–86.

Russ, J. C. 1989. Computerized object recognition using contextual learning. *J. Comput. Assist. Microsc.* 1: 105–129.

Russ, J. C. 1990. *Computer-Assisted Microscopy. The Measurement and Analysis of Images.* Plenum Press, New York.

Ruzicka, J., P. J. Baxter, O. Thastrup, and K. Scudder. 1996. Flow injection microscopy: a novel tool for the study of cellular response and drug discovery. *Analyst.* 121: 945–950.

Samekh, M. 1990. Image formation in scanned heterodyne microscope systems. *J. Microsc.* 160: 225–243.

Schinagl, R. M., M. K. Ting, J. H. Price, and R. L. Sah. 1996. Video microscopy to quantitate the inhomogenous equilibrium strain within articular cartilage during confined compression. *Ann. Biomed. Eng.* 24: 500–512.

Schorsch, C., C. Garnier, and J. L Doublier. 1995. Microscopy of xanthan/galactomannan mixtures. *Carbohydr. Polym.* 28: 319–323.

Seitz, W. R. 1984. Chemical sensors based on fiber optics. *Anal. Chem.* 56: 17A–34A.

Shackleford, J. M. and K. L. Yielding. 1987. Application of the fiber-optic perfusion fluorometer to absorption and exsorption studies in hairless mouse skin. *Anat. Rec.* 219: 102–107.

Shaw, P. J. 1990. Three-dimensional optical microscopy using tilted views. *J. Microsc.* 158: 165–172.

Shotton, D. M. 1991. Video and opto-digital imaging microscopy. In: R. C. Cherry (Ed.), *New Techniques of Optical Microscopy and Microspectrophotometry.* CRC Press, Boca Raton, FL, pp. 1–47.

Slayter, E. M. 1970. *Optical Methods in Biology*. Wiley-Interscience, New York, pp. 583–584.

Stamm, W., K. Sauerland, and N. Muller. 1993. A low cost high-voltage controller for photomultiplier tubes. *Nucl. Instrum. Methods Phys. Res. A, Accel. Spectrom. Detect. Assoc. Equip. (Neth.).* A328: 601–602.

Stavenga, D. G. and H. L. Leertouwer. 1990. Curvature measurement with reflected-light microscopy and its application to fly facet lenses. *J. Microsc.* 158: 87–93.

Steinhart, H., A. Bosselmann, and C. Möller. 1994. Determination of pyridinolines in bovine collagenous tissues. *J. Agric. Food Chem.* 42: 1943–1947.

Stevenson, R. 1996. Bioapplications and instrumentation for light microscopy in the 1990s. *Am. Lab.* 28(6): 28–51.

Suyama, K. and F. Nakamura. 1992. Two fluorescent crosslinking amino acids having N-substituted dihydrooxopyridine skeleton isolated from bovine elastin. *Bioorg. Med. Chem. Lett.* 2: 1767–1770.

Swatland, H. J. 1979. Endomysial boundary scanning as a method of counting skeletal muscle fibers. *Mikroskopie.* 35: 280–288.

Swatland, H. J. 1984a. The radial distribution of succinate dehydrogenase activity in skeletal muscle fibres. *Histochem. J.* 16: 321–329.

Swatland, H. J. 1984b. Intracellular distribution of succinate dehydrogenase activity in skeletal muscle fibers of geese. *Can. J. Zool.* 62: 235–240.

Swatland, H. J. 1985a. Growth-related changes in the intracellular distribution of succinate dehydrogenase in turkey muscle. *Growth.* 49: 409–416.

Swatland, H. J. 1985b. Patterns of succinate dehydrogenase activity in a leg muscle of the domestic duck during post-natal development. *Can. J. Zool.* 63: 55–57.

Swatland, H. J. 1987. Fluorimetry of bovine myotendon junction by fibre-optics and microscopy of intact and sectioned tissues. *Histochem. J.* 19: 276–280.

Swatland, H. J. 1988. Autofluorescence of bovine ligamentum nuchae, cartilage, heart valve and lung measured by microscopy and fibre-optics. *Res. Vet. Sci.* 45: 230–233.

Swatland, H. J. 1989a. Thermal denaturation of perimysial collagen in meat measured by polarized light microscopy. *J. Food Sci.* 55: 305–311.

Swatland, H. J. 1989b. Birefringence of beef and pork muscle fibers measured by scanning and ellipsometry with a computer-assisted polarising microscope. *J. Comput. Assist. Microsc.* 1: 249–262.

Swatland, H. J. 1990a. Intracellular glycogen distribution examined interactively with a light microscope scanning stage. *J. Comput. Assist. Microsc.* 2: 233–237.

Swatland, H. J. 1990b. Questions in programming a fluorescence microscope. *J. Comput. Assist. Microsc.* 2: 125–132.

Swatland, H. J. 1993. Photomultiplier response in a computer-assisted microscope photometer measured with a multiprogrammer. *J. Comput. Assist. Microsc.* 5: 231–235.

Swatland, H. J. 1994. Program control of changes in the pH of biological specimens for light microscopy. *J. Comput. Assist. Microsc.* 6: 41–46.

Swatland, H. J. 1995. Surface reflectance of meat measured by microscope polarimetry and spectrophotometry using a tilting stage with lateral illumination. *J. Comput. Assist. Microsc.* 7: 211–219.

Swatland, H. J. 1996a. Monitoring photomultiplier performance in a computer-assisted light microscope. *J. Comput. Assist. Microsc.* 8: 75–82.

Swatland, H. J. 1996b. Effect of stretching pre-rigor muscle on the back-scattering of polarized near-infrared. *Food Res. Int.* 29: 445–449.

Swatland, H. J., T. C. Irving, and B. M. Millman. 1989. Fluid distribution in pork, measured by X-ray diffraction, interference microscopy and centrifugation compared to paleness measured by fiber optics. *J. Anim. Sci.* 67: 1465–1470.

Swift, H. and E. Rasch. 1956. Microphotometry with visible light. In: G. Oster and A. W. Pollister (Eds.), *Physical Techniques in Biological Research*, Vol. III. Academic Press, New York, pp. 353–400.

Tas, J., M. van der Ploeg, J. P. Mitchell, and N. S. Cohn. 1980. Protein staining methods in quantitative cytochemistry. *J. Microsc.* 119: 295–311.

Taylor, D. L. 1975. Birefringence changes in vertebrate striated muscle. *J. Supramolec. Struct.* 3: 181–191.

Theorell, H. and C. de Duve. 1947. Crystalline human myoglobin from heart muscle and urine. *Arch. Biochem.* 12: 113–124.

Thetford, A. and S. C. Simmens. 1969. Birefringence phenomena in cylindrical fibres. *J. Microsc.* 89: 143–150.

Thomas, C., P. Devries, J. Hardin, and J. White. 1996. Four-dimensional imaging: computer visualization of 3D movements in living specimens. *Science.* 273: 603–607.

Tian, R. and M. A. J. Rodgers. 1991. Time-resolved fluorescence microscopy. In: R. C. Cherry (Ed.), *New Techniques of Optical Microscopy and Microspectrophotometry.* CRC Press, Boca Raton, FL, pp. 177–198.

Tiffe, H-W. and H. Hundeshagen. 1982. Investigation of fading and recovery of fluorescent intensity at 73.5 K. *J. Microsc.* 126: 231–235.

Tomasek, J. J., S. W. Meyers, J. B. Basinger, D. T. Green, and R. L. Shew. 1994. Diabetic and age-related enhancement of collagen-linked fluorescence in cortical bones of rats. *Life Sci.* 55: 855–861.

Tomita, M., Y. Fukuuchi, N. Tanahashi, M. Kobari, Y. Terayama, T. Shinohara, S. Konno, H. Takeda, D. Itoh, M. Yokoyama, S. Terakawa, and H. Haapaniemi. 1995. Activated leukocytes, endothelial cells, and effects of pentoxifylline: observations by VEC-DIC microscopy. *J. Cardiovasc. Pharmacol.* 25: S34–39.

Tonna, E. A. and E. J. Rogers. 1968. Microscopic photodensitometry and microspectrophotometry using a fixed double-aperture optical system and an electronically controlled automatic scanning microscope stage. *J. Roy. Microsc. Soc.* 88: 71–84.

Tsay, T-T., R. Inman, B. Wray, B. Herman, and K. Jacobson. 1990. Characterisation of low-light-level cameras for digitized video microscopy. *J. Microsc.* 160: 141–159.

Udenfriend, S. 1969. Fluorescence as a probe into mechanisms in intact cells and sub-cellular elements. In: J. D. Winefordner, P. A. St. John, and W. J. McCarthy (Eds.), *Fluorescence Assay in Biology and Medicine,* Vol. II. Academic Press, New York, pp. 539–566.

van Aspert van Erp, A. J. M., A. E. van't Hof Grootenboer, G. Brugal, and G. P. Vooijs. 1996. Individual use of cytomorphologic characteristics in the diagnosis of endocervical columnar cell abnormalities: selection of preferred features with help of the NAVIGATOR microscope. *Anal. Cell. Pathol.* 11: 73–95.

Van Noorden, C. J. F. and J. Tas. 1981. Model film studies in enzyme histochemistry with special reference to glucose-6-phosphate dehydrogenase. *Histochem. J.* 13: 187–206.

Velapoldi, R. A., J. C. Travis, W. A. Cassatt, and W. T. Yap. 1974. Inorganic ion-doped glass fibres as microspectrofluorimetric standards. *J. Microsc.* 103: 293–303.

Vergara, J., M. DiFranco, D. Compagnon, and B. A. Suarez-Isla. 1991. Imaging of calcium transients in skeletal muscle fibers. *Biophys. J.* 59: 12–24.

Visser, J. W. M., A. A. M. Jongeling, and H. J. Tanke. 1979. Intracellular pH-determination by fluorescence measurements. *J. Histochem. Cytochem.* 27: 32–35.

Vollath, D. 1987. Automatic focusing by correlative methods. *J. Microsc.* 147: 279–288.

Waggoner, A., L. Taylor, A. Seadler, and T. Dunlay. 1996. Multiparameter fluorescence imaging microscopy: reagents and instruments. *Human Pathol.* 27: 494–502.

Walker, P. J. and J. M. A. Watts. 1970. Permanent fluorescent test slides. *J. Microsc.* 92: 63–65.

Ward, E. H. and C. L. Hussey. 1987. Remote acquisition of spectroelectrochemical data in a room-temperature ionic liquid with a microprocessor-controlled fiber-optic spectrophotometry system. *Anal. Chem.* 59: 213–217.

Weidner, V. R. and J. J. Hsia. 1981. Reflection properties of pressed polytetrafluoroethylene powder. *J. Opt. Soc. Am.* 71: 856–861.

Westerblad, H. and J. Lännergren. 1990. Reversible increase in light scattering during recovery from fatigue in *Xenopus* muscle fibres. *Acta Physiol. Scand.* 140: 429–435.

Wharton, D. A. and J. J. Roland. 1984. A thermoelectric microscope stage for the measurement of the supercooling points of microscopic organisms. *J. Microsc.* 134: 299–305.

Whittaker, P., R. A. Kloner, D. R. Boughner, and J. G. Pickering. 1994. Quantitative assessment of myocardial collagen with picrosirius red staining and circularly polarized light. *Basic Res. Cardiol.* 89: 397–410.

Whittaker, P., M. E. Schwab, and P. B. Canham. 1988. The molecular organization of collagen in saccular aneurysms assessed by polarized light microscopy. *Connective Tiss. Res.* 17: 43–54.

Williams, B. E., D. M. Binkley, and M. E. Casey. 1992. A remote gain control for photomultiplier tubes. *IEEE Nuclear Science Symposium and Medical Imaging Conference*, Vol. 1. Orlando, FL, pp. 245–247.

Wolman, M. and T. Gillman. 1972. A polarized light study of collagen in dermal wound healing. *Br. J. Exp. Pathol.* 53: 85–89.

Wood, J. R. and D. A. I. Goring. 1974. Ultraviolet microscopy at wavelengths below 240 nm. *J. Microsc.* 100: 105–111.

Wreford, N. G. M. and G. C. Schofield. 1975. A microspectrofluorometer with on-line real time correction of spectra. *J. Microsc.* 103: 127–130.

Yang, T-T., S. R. Kain, P. Kitts, A. Kondepudi, M. M. Yang, and D. C. Youvan. 1996. Dual color microscopic imagery of cells expressing the green fluorescent protein and a red-shifted variant. *Gene* 173: 19–23.

Yao, Y. J. and S. F. Y. Li. 1996. Determination of erythrocyte porphyrins by epi-illumination fluorescence microscope with capillary electrophoresis. *J. Liq. Chromat. Rel. Technol.* 19: 1–15.

Yeh, Y., R. J. Baskin, K. Burton, and J. S. Chen. 1987. Optical ellipsometry on the diffraction order of skinned fibers: pH-induced rigor effects. *Biophys. J.* 51: 439–447.

Zeiss, 1980. *Determination of Relative Spectral Fluorescence Intensities,* Bull. A41-823.2-e. Carl Zeiss, Oberkochen, Germany.

Zimmer, H-G. 1979. Digital picture analysis in microphotometry of biological materials. *J. Microsc.* 116: 365–372.

Zubchenok, V. Y., A. A. Nakhan'kov, I. V. Reznikov, and E. D. Tverdokhlebov. 1988. General-purpose programmable voltage divider for photomultipliers with increased loads. *Instrum. Exp. Tech. (U.S.).* 30: 367–370.

Index

fluorspar, 137
focus, 11–12, 129
format, 42
Foster prism, 118
fovea, 115
frame-grabber, 101, 153
Fresnel, 196

G

gene expression 138
glare, 7, 14, 35
gloss, 123, 133
glycogen, 110–111
goniospectrophotometry, 187–188
GPIB, 40, 212
graded index, 177
graphics, 25–26, 71–74
grating efficiency, 46
gray level, 154
gray–map, 110
ground, 21

H

half-step, 103
haloes, 16
handshaking, 22–23
HBO, *see* mercury arc
heart failure, 189
    valve, 137
heat filter, 14, 141
heavy metal, 142
histogram, 171–173
HP-IB, *see* GPIB
hypodermic, 27, 198

I

I/O, 19
IEEE 488, *see* GPIB
illumination, 33, 127–129
illuminator, 29–34
    adjustments, 3–4
    assembly, 4–5

centering, 6–7
IMAGE, 106
immersion oil, 10
immunofluorescence, 137
interfacing, 19–25
interference, 15–16, 94, 115
interferometry, 17
interrupt, 20
iridescence, 134
iris, 13, 18

J

joystick, 105

K

keratin, 137
kernel, Laplacian, 162
    neighborhood averaging, 159–160
Köhler illumination, 5–8, 29, 93

L

laser, 30, 168, 185
light guide, 178–179
lignin, 137
lipofuscin, 143
listener, 19
liver, 191
LOCAL, 105
LOCAL LOCKOUT, 20
lung, 137, 143

M

magnesium oxide, 181
Makler, 141
mapping, concentric, 111
    histochemical, 107–112
    software, 108–110
memory, 42
mercury arc, 30
methyl green, 83
Michel-Lévy chart, 115